SpringerBriefs in Computer Science

Series Editors

Stan Zdonik, Providence, USA
Peng Ning, Raleigh, USA
Shashi Shekhar, Minneapolis, USA
Jonathan Katz, College Park, USA
Xindong Wu, Burlington, USA
Lakhmi C. Jain, Adelaide, Australia
David Padua, Urbana, USA
Xuemin Shen, Waterloo, Canada
Borko Furht, Boca Raton, USA
V. S. Subrahmanian, College Park, USA
Martial Hebert, Pittsburgh, USA
Katsushi Ikeuchi, Tokyo, Japan
Bruno Siciliano, Napoli, Italy

T0213906

For further volumes:
http://www.springer.com/series/10028

Marcos R. Vieira · Vassilis J. Tsotras

Spatio-Temporal Databases

Complex Motion Pattern Queries

 Springer

Marcos R. Vieira
IBM Research Laboratory Brazil
Rio de Janeiro
Brazil

Vassilis J. Tsotras
Department of Computer Science
and Engineering
Bourns College of Engineering
University of California
Riverside, CA
USA

ISSN 2191-5768   ISSN 2191-5776 (electronic)
ISBN 978-3-319-02407-3 ISBN 978-3-319-02408-0 (eBook)
DOI 10.1007/978-3-319-02408-0
Springer Cham Heidelberg New York Dordrecht London

Library of Congress Control Number: 2013949410

Printed on acid-free paper

Springer is part of Springer Science+Business Media (www.springer.com)

Preface

This book presents several novel query processing techniques, called complex motion pattern queries, specifically designed for very large *Spatio-Temporal Databases* of moving objects (also called trajectories). First, it begins with the definition of *flexible pattern query*, a very powerful, yet easy to use motion pattern query that allows users to select trajectories based on specific events of interest. *Flexible pattern query* is described as regular expressions over a spatial alphabet that can be implicitly/explicitly "anchored" to the time domain. Moreover, it allows users to include *variables* in the pattern query, and thus greatly increase its expressive power. Second, the *Spatio-Temporal Pattern System* (STPS) is presented, which is an adaptation of *flexible pattern query* for large mobile phone-call databases. These databases contain many millions of records with information about mobile phone calls, including the user's location, when the call was made/received, and duration of the call, among other useful information. The design of STPS takes into consideration the layout of the areas being covered by the cellular towers, as well as *areas* that label places of interest (e.g., neighborhoods, airports, parks). Third, *density-based pattern query* is described for discovering trajectories that follow a pattern that captures the *aggregate* behavior of trajectories as groups. Several evaluation algorithms are presented for finding groups of trajectories that move *together* in space and time, i.e., within a predefined distance to each other for a continuous period of time. The last complex motion pattern query proposed in this book is for diversifying query results. The goal of this query is to build a result that contains *relevant* elements to the user's query and *diverse* elements to other elements in the result. This pattern query is useful when, for example, an exploratory query to a very large database leads to a vast number of answers in the result. Navigating through such a large result requires huge effort and users give up after perusing through the first few results, thus some interesting results hidden further down the result set can easily be missed. To overcome this problem, a generic framework, called DivDB, for diversifying query results is presented. Two

new evaluation methods, as well as several existing ones, are described and tested
in the proposed DivDB framework. The efficiency and effectiveness of all the
proposed complex motion pattern queries are demonstrated through an extensive
experimental evaluation using real and synthetic spatio-temporal databases.

Rio de Janeiro, RJ, Brazil, August 2013   Marcos R. Vieira
Riverside, CA, USA   Vassilis J. Tsotras

Acknowledgments

This material is based upon work supported by National Science Foundation (NSF) under grants IIS-0534781, IIS-0803410, IIS-0910859 and IIS-0705916, CAPES (Brazilian Federal Agency for Post-Graduate Education), and the Fulbright Program.

The authors also thank the following collaborators for providing valuable support for the development of this work: Petko Bakalov, Maria Camila Barioni, Michalis Faloutsos, Enrique Frías-Martínez, Vanessa Frías-Martínez, Marios Hadjieleftheriou, Erik Hoel, Eamonn Keogh, Walid Najjar, Nuria Oliver, Humberto Razente, Divesh Srivastava, Caetano Traina Jr, and Neal Young.

Contents

Acronyms

BFE	Basic Flock Evaluation Algorithm
BTS	Base Transceiver Stations
CDR	Call Detail Record
CLT	Clustering-Based Method
CRE	Continuous Refinement Evaluation Algorithm
DPP	Dynamic Programming Pattern Algorithm
E-KMP	Extended-Knuth-Morris-Pratt Algorithm
E-NFA	Extended-Nondeterministic Finite Automaton Algorithm
FPGA	Field Programmable Gate Arrays
GMC	Greedy with Marginal Contribution Method
GNE	Greedy-Randomized with Neighborhood Expansion Method
GPS	Global Positioning System
GPU	Graphics Processing Units
GRASP	Greedy-Randomized Adaptive Search Procedure
IJP	Index Join Pattern Algorithm
KMP	Knuth-Morris-Pratt
LCS	Longest Common Subsequence
MBR	Minimum Bounding Regions
MMC	Maximum Marginal Contribution
MMR	Maximal Marginal Relevance Method
MMS	Multimedia Messaging Service
MSD	Max-Sum Dispersion Method
NFA	Nondeterministic Finite Automaton
NN	Nearest Neighbor
PFE	Pipe Filtering Evaluation Algorithm
RDBMS	Relational Database Management System
RCL	Restricted Candidate List
RFID	Radio-Frequency Identification
SMS	Sort Message Service
STPS	Spatio-Temporal Pattern System
TA	Threshold Algorithm
TDE	Top-Down Evaluation Algorithm

Chapter 1
Introduction

The wide availability and use of location and mobile technologies (e.g. cheap GPS devices, ubiquitous cellular networks, RFID (Radio-Frequency Identification) tags), has enabled many applications that generate and maintain large amounts of data in the form of trajectories (see [14] for a survey on trajectory data). For example, new generations of monitoring/tracking systems have emerged, providing complex services to their end users. The quality and accuracy of the produced spatio-temporal data has also improved: instead of the traditional cell phone tower triangulation method, assisted GPS (A-GPS) [5] was recently introduced to improve location accuracy (such as enhanced 911 service [3]). These advances have led to the creation of large volumes of accurate spatio-temporal data in the form of trajectories. A trajectory has a unique identifier and contains location data (e.g. latitude/longitude) gathered for a specific moving object over an ordered sequence of time instants (or intervals). Given the huge volume of data generated in the form of trajectories, there is a need to develop better and more efficient techniques for data management and query evaluation over trajectories.

Past research efforts on querying trajectory data has mainly concentrated on traditional spatio-temporal queries, such as range and nearest neighbors searches (e.g. finding trajectories that passed by a predefined area during a certain period of time, see Fig. 1.1a), or similarity/clustering based tasks, such as extracting similar movement patterns and periodicities from trajectory data (e.g. finding all trajectories that are similar to a given query trajectory according to a predefined similarity measure, see Fig. 1.1b). The main problems with the above two approaches are that a range query may retrieve too many results (as exemplified in Fig. 1.1a), while a similarity/clustering based task may be too restrictive and, thus, return no result (as illustrated in Fig. 1.1b).

Nevertheless, trajectories are complex objects whose behavior over space and time can be better captured as a *sequence* of interesting events defined by the user. Additionally, trajectories can be searched by the *aggregate* behavior of trajectories as groups interacting in space and time. These two types of patterns for querying

M. R. Vieira and V. J. Tsotras, *Spatio-Temporal Databases*,
SpringerBriefs in Computer Science, DOI: 10.1007/978-3-319-02408-0_1,
© The Author(s) 2013

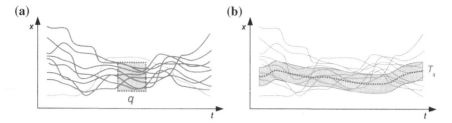

Fig. 1.1 (a) Trajectory range query. (b) Trajectory similarity-based query. Examples of trajectory-based queries and their problems

trajectories are called "motion patterns queries" [13], which are discussed in more detail in this book.

In this book, we first introduce *flexible pattern queries* [7, 8], which allow users to select trajectories based on specific events of interest. Figure 1.2a illustrates a flexible pattern query defined by a sequence of 3 spatio-temporal predicates. Such pattern queries are described as regular expressions over a spatial alphabet that can be implicitly or explicitly "anchored" to the time domain. Moreover, flexible pattern query allows users to include *variables* in the query, and thus greatly increase its expressive power. We describe our proposed framework, called FlexTrack [8], for efficient processing of flexible pattern queries on trajectorial databases.

Next, we present the Spatio-Temporal Pattern System (STPS) for querying spatio-temporal patterns in mobile phone-call databases [9]. In this work, we adopt the FlexTrack system and study its application in the domain of Call Detail Record (CDR) databases. The STPS allows users to express mobility pattern queries with a regular expression-like language. The join-based evaluation algorithm proposed in [7] was modified to handle trajectories specified in CDR format rather than the traditional form. This change in the data format also poses changes in the FlexTrack query language. The STPS query evaluation system is redesigned to work with the features (e.g. relational tables, B^+-trees) of a commercially available Relational Database Management System (RDBMS), since CDR databases are typically implemented in such systems.

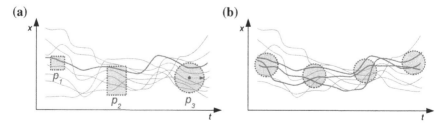

Fig. 1.2 (a) Flexible pattern query with 3 spatio-temporal predicates. (b) Density-based pattern query. Examples of motion patterns queries

We then turn our attention to *density-based pattern queries* [6, 10], which search for trajectories that follow a pattern that captures the *aggregate* behavior of trajectories as groups. Such queries discover groups of moving objects that have a "strong" relationship in the space for a given time duration. Figure 1.2b shows a density-based pattern query defined by a group of two trajectories that move together for at least four timestamps. In this work we consider the discovery of *flock* patterns among the moving objects, i.e., the problem of identifying all groups of trajectories that stay "together" in space for the duration of a given time interval. We consider moving objects to be "close" together if there exists a defined *disk* with a given radius that covers all moving objects in the pattern. A trajectory satisfies the above pattern as long as "enough" other trajectories are contained inside the disk for the specified time interval; that is, the answer is based not only on a given trajectory's behavior but also on the ones near it. Consider, for example, finding groups of trajectories that move "together", i.e. within a predefined distance to each other, for a certain continuous period of time. Such queries typically arise in surveillance applications, e.g., identify groups of suspicious people, convoys of vehicles, migration of birds, flocks of animals [1, 2, 4].

In all of the previously described types of pattern queries, all the elements in the answer set are *relevant* to the user's pattern query. Nevertheless, with the availability of very large databases, an exploratory query can easily lead to a vast answer set, typically based on an answer's *relevance* to the user query. Navigating through such an answer set requires huge effort and users give up after perusing through the first few answers, thus some interesting answers hidden further down the answer set can easily be missed. An approach to address this problem is to present the user with the most *diverse* among the answers based on some diversity criterion. To cope with this problem, we propose the DivDB framework, a general framework for evaluation and optimization of methods for diversifying query results [11, 12]. In the DivDB framework, an initial ranking candidate set produced by a query is used to construct a result set, where elements are ranked with respect to *relevance* and *diversity* features, i.e., the retrieved elements should be as relevant as possible to the query, and, at the same time, the result set should be as diverse as possible. While addressing relevance is relatively simple and has been heavily studied, diversity is a harder problem to solve. One major contribution of this work is that several existing methods for diversifying query results were adapted, implemented and evaluated in the DivDB framework. We also propose two new approaches, namely the Greedy with Marginal Contribution (GMC) and the Greedy Randomized with Neighborhood Expansion (GNE) methods. Both methods iteratively construct a result set using a scoring function that ranks candidate elements using not only relevance and diversity to the existing result set, but also accounts for diversity against the remaining candidates.

The rest of this book is organized as follows: Chap. 2 presents flexible pattern queries; Chap. 3 describes the Spatio-Temporal Pattern System (STPS) for mobile phone-call databases; Chap. 4 discusses *density-based pattern queries*; Chap. 5 proposes the DivDB framework for diversifying query results; Finally, Chap. 6 concludes this book.

References

1. Benkert, M., Gudmundsson, J., Hübner, F., Wolle, T.: Reporting flock patterns. In: Proceedings of the Conference on Annual European Symposium on Algorithms (ESA), pp. 660–671. Springer-Verlag (2006). DOI http://dx.doi.org/10.1007/11841036_59
2. Benkert, M., Gudmundsson, J., Hübner, F., Wolle, T.: Reporting flock patterns. Comput. Geom. Theory Appl. **41**, 111–125 (2008). DOI http://dx.doi.org/10.1016/j.comgeo.2007.10.003
3. Consumer & Governmental Affairs Bureau, F.: Wireless 911 services. http://www.fcc.gov/guides/wireless-911-services
4. Gudmundsson, J., van Kreveld, M.: Computing longest duration flocks in trajectory data. In: Proceedings of the ACM SIGSPATIAL International Conference on Advances in Geographic Information Systems, pp. 35–42. ACM (2006). http://dx.doi.org/10.1145/1183471.1183479
5. NAVCEN, U.C.G.N.C.: Navstar GPS User Equipment Introduction. DOI http://www.navcen.uscg.gov/pubs/gps/gpsuser/gpsuser.pdf(1996)
6. Vieira, M.R., Bakalov, P., Tsotras, V.J.: On-line discovery of flock patterns in spatio-temporal data. In: Proceedings of the ACM SIGSPATIAL International Conference on Advances in Geographic Information Systems, pp. 286–295. ACM (2009). DOI http://dx.doi.org/10.1145/1653771.1653812
7. Vieira, M.R., Bakalov, P., Tsotras, V.J.: Querying trajectories using flexible patterns. In: Proceedings of the International Conference on Extending Database Technology (EDBT), pp. 406–417. ACM (2010). DOI http://dx.doi.org/10.1145/1739041.1739091
8. Vieira, M.R., Bakalov, P., Tsotras, V.J.: FlexTrack: A system for querying flexible patterns in trajectory databases. In: Proceedings of the International Symposium on Advances in Spatial and Temporal Databases (SSTD), *Lecture Notes in Computer Science*, vol. 6849, pp. 475–480. Springer Berlin Heidelberg (2011). DOI http://dx.doi.org/10.1007/978-3-642-22922-0_34
9. Vieira, M.R., Frias-Martinez, E., Bakalov, P., Frias-Martinez, V., Tsotras, V.J.: Querying spatio-temporal patterns in mobile phone-call databases. In: Proceedings of the IEEE International Conference on Mobile Data Management (MDM), pp. 239–248. IEEE Computer Society (2010). DOI http://dx.doi.org/10.1109/MDM.2010.24
10. Vieira, M.R., Frias-Martinez, E., Oliver, N., Frias-Martinez, V.: Characterizing dense urban areas from mobile phone-call data: Discovery and social dynamics. In: Proceedings of the IEEE International Conference on Social Computing (SocialCom), pp. 241–248. IEEE Computer Society (2010). DOI http://dx.doi.org/10.1109/SocialCom.2010.41
11. Vieira, M.R., Razente, H., Barioni, M.C., Hadjieleftheriou, M., Srivastava, D., Jr., C.T., Tsotras, V.J.: DivDB: A system for diversifying query results. Proceedings of the VLDB Endowment (PVLDB) **4**(12), 1395–1398 (2011)
12. Vieira, M.R., Razente, H., Barioni, M.C., Hadjieleftheriou, M., Srivastava, D., Jr., C.T., Tsotras, V.J.: On query result diversification. In: Proceedings of the IEEE International Conference on Data Engineering (ICDE), pp. 1163–1174. IEEE Computer Society (2011). DOI http://dx.doi.org/10.1109/ICDE.2011.5767846
13. Vieira, M.R., Tsotras, V.J.: Complex motion pattern queries for trajectories. In: Proceedings of the IEEE International Conference on Data Engineering (ICDE) Workshops, pp. 280–283. IEEE Computer Society (2011). DOI http://dx.doi.org/10.1109/ICDEW.2011.5767665
14. Zheng, Y., Zhou, X. (eds.): Computing with Spatial Trajectories. Springer (2011). DOI http://dx.doi.org/10.1007/978-1-4614-1629-6

Chapter 2
Flexible Pattern Queries

2.1 Introduction

The wide availability of location and mobile technologies (e.g., cheap GPS devices, ubiquitous cellular networks, RFID tags) as well as the improved location accuracy (e.g., *A-GPS* [1], E-911 [2]) has enabled a wide variety of applications that generate and maintain data in the form of *trajectories*. Examples include *AccuTracking* [3], *tracNET24* [4], Path Intelligence's *FootPath* [5], InSTEDD's *GeoChat* [6], Geo-fencing [7], among many others. In such applications, each trajectory has a unique identifier and consists of location data gathered for a specific moving object over an ordered sequence of time instants. Given the high data volume generated by the above new applications, more effective and efficient techniques for query evaluation over trajectory data are needed.

Previous work on querying trajectories can be divided in two main categories: (**a**) querying the future movements of moving objects based on their current positions (e.g., [8–12]); and (**b**) querying trajectory archives, which is also the focus of this chapter. Recent research efforts on querying trajectory archives has concentrated on (**i**) traditional spatio-temporal queries, such as Range, Nearest Neighbor (*NN*), and Reverse Nearest Neighbor (*RNN*) searches (e.g., [13–18]) (e.g., finding all trajectories that passed by *Lower Manhattan* at *10:30 am*), or (**ii**) similarity/clustering based tasks (e.g., [19–24]), such as extracting similar movement patterns and periodicities from a trajectory archive (e.g., finding all trajectories in the archive that are similar to a given query trajectory according to some similarity measure).

However, given the nature of trajectories as typically long sequences of spatio-temporal events, a single range predicate query may provide too many results (e.g., many trajectories passed through *Midtown Manhattan*, as illustrated in Fig. 2.1a) while a similarity-based query may be too restrictive (e.g., not many trajectories match the full extent or large part of the query trajectory, as shown in Fig. 2.1b). To overcome the above drawbacks, we thus propose a different approach of using *motion pattern queries*. A motion pattern query specifies a combination of predicates that can thus capture only the parts of the trajectories that are of interest to the user.

M. R. Vieira and V. J. Tsotras, *Spatio-Temporal Databases*,
SpringerBriefs in Computer Science, DOI: 10.1007/978-3-319-02408-0_2,

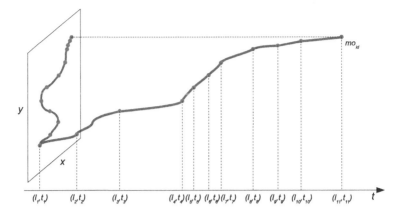

Fig. 2.1 Example of a trajectory for mo_{id} with a sequence of 11 locations

For example: "find trajectories that first went by *Midtown Manhattan*, later went by *Queens* and then ended up in *Brooklyn*". This query simply provides a sequence of range predicates that have to be satisfied in the specified order (i.e., first *Midtown Manhattan*, then *Queens*, and then *Brooklyn*). One can also add *NN* predicates as well, For example, in the previous query: "... and they were closest to *the JFK airport*" as well as explicit time constraints: "... ended up in *Brooklyn* between 1 and 4 pm". Conceptually, motion pattern queries cover the query choices between the above two extremes: single predicates and similarity queries.

In this chapter, we introduce FlexTrack [25, 26], a general and powerful system that defines pattern queries as regular expressions over a finite spatial alphabet. Each letter in the alphabet corresponds to a non-overlapping region in the spatial domain; the union of all regions covers the entire space domain where the trajectories in the database lie. We note that there are various advantages from these choices:

1. The use of non-overlapping regions is natural: trajectories correspond to real entities and hence a trajectory can be in a single region in the spatial domain at a given time;
2. Raw trajectory data typically come, for example, from sensors, GPS, RFID readers, etc., and provide extra detail that becomes cumbersome to query. Instead, the regions offer a very user friendly way to express queries since the user is more familiar with the spatial regions [27] (e.g., Lower Manhattan, JFK airport, Brooklyn bridge);
3. The use of spatial regions allows high-level summarization and filtering of trajectories in the database. The region description of a trajectory is much compact, leading to faster query processing, while the raw data is still kept if more detail is needed;
4. Spatial regions enable easy and effective indexing; it further enables the use of alternative evaluation algorithms (e.g., joins among inverted indexes, pattern matching, NFAs);

5. The region alphabet facilitates querying by regular expressions as a query language: the user can now describe complex queries over paths using the fixed alphabet.

The FlexTrack system uses a hierarchical region alphabet, where the user has the ability to define queries with finer alphabet granularity (*zoom in*) for the portions of greater interest, and higher granularity (*zoom out*) elsewhere. Through a GUI the user specifies a pattern query by selecting regions using various levels of the hierarchy; the user's query is then automatically translated into a regular expression over the finest region granularity to be further executed. This chapter describes how queries are executed at the finer granularity in the FlexTrack system. Hence we assume that all query regions are at the same (finer) granularity. This finer granularity is chosen by and depends on the application needs.

The described FlexTrack system further allows the use of *variables* within the pattern query. We term these *variable*-enabled queries as "flexible" patterns, as they lead to a very powerful way to query the trajectory database. Moreover, the *fixed* and/or *variable* spatial predicates can express explicit temporal constraints ("between 1 and 4 pm") and/or implicit temporal ordering between them ("anytime later"). Flexible pattern queries can also include "numerical" conditions (*NN* and their variants) over the duration of the trajectory. Using the FlexTrack system user can "focus" only on the portions/events in a trajectory's lifetime that are of his/her interest.

The FlexTrack has two novel algorithms to efficiently process such complex pattern queries over very large trajectory databases. These two query evaluation algorithms first concentrate on trajectories that satisfy *all* the fixed predicates specified in the pattern query. As such, both methods effectively prune large portions of the database that cannot lead to query answers. The first proposed algorithm uses the merge-join paradigm over the lists of trajectories associated with the query predicates. The second algorithm is based on a dynamic programming technique that finds subsequence matches between the trajectory region representations and the pattern query. The second phase of both algorithms then evaluate of the remaining *variable* predicates.

We should emphasize that pattern as effective ways to query databases have been examined in the past. For instance, [28–30] examine, respectively, patterns over time and event streams. Trajectory databases differ for those domains mainly because trajectories have both spatial and temporal behavior, which make the problem more complex to handle. There are also some works [31–34] that propose the use of patterns in spatio-temporal databases, as detailed in the next section. However, previous approaches either concentrate on language/modeling related issues, provide less query support (e.g., no temporal and/or numerical constraints), or have less efficient/general evaluation methods than the FlexTrack system.

To summarize, this chapter describes the FlexTrack system with the following main contributions:

1. It is a simple yet powerful system using a query language based on regular expressions;
2. It allows patterns to contain variables over the query space regions;

3. It contains two efficient evaluation algorithms that use lightweight index structures, which can be easily implemented in most commercial DBMS nowadays;
4. We present an extensive experimental evaluation of the two proposed methods against other improved methods: a NFA-based method (Nondeterministic Finite Automaton) [30] and a *KMP*-based method (Knuth-Morris-Pratt) [31].

The remainder of this chapter is organized as follows: Section 2.2 discusses the related work; Sect. 2.3 provides the basic definitions and formal description of the spatio-temporal query language; The proposed FlexTrack system is described in details in Sect. 2.4 and its experimental evaluation appears in Sects. 2.5; Sect. 2.6 concludes the chapter with the final remarks.

2.2 Related Work

Previous works on evaluating spatio-temporal queries have mainly focused on either the evaluation of a single spatio-temporal predicate (e.g., Range or Nearest Neighbor *NN* queries), or similarity/clustering queries where the entire (or a part of) trajectory is given as query and similar trajectories are sought.

Queries with a single spatio-temporal predicate for trajectory databases have been extensively studied in the past [14–17, 35]. To make the evaluation process more efficient, the query predicates are typically evaluated utilizing hierarchical spatio-temporal indexing structures [9, 13, 36–42]. Most of these indexing structures use the concept of Minimum Bounding Regions (MBR) to approximate trajectories, which are then indexed using traditional spatial access methods, like the MVR-tree [43] and TPR-tree [44]. These solutions, however, are focused only on single spatio-temporal predicate queries. Thus, none of them can be used for efficient evaluation of flexible pattern queries with multiple predicates. Moreover, the FlexTrack system is different than (and orthogonal to) approaches like [45] that can handle a set of single and independent predicates (i.e., different queries). In complex pattern queries all predicates appear in the same pattern query, and should all be satisfied by each trajectory in the result set.

Searching for trajectories that are similar to a given query trajectory has also been well studied in the past [46–50]. Basically, the problem is to search for all trajectories in the database similar to a given threshold value to a query trajectory that is represented by an entire, or a part of, trajectory. Works in this area concentrate mainly on the use of different distance metrics to measure the similarity between trajectories. For example, non-metric similarity functions based on the Longest Common Subsequence (LCS) are examined in [51]. Cai and Ng [47] proposes to approximate and index a multidimensional spatio-temporal trajectory with a low order continuous Chebyshev polynomial which can then lead to efficient indexing for similarity queries [48].

The importance of pattern queries has been recognized in the relational databases domain as well. For instance, Informix [52] introduced a time-series database and provided a library for pattern searching that can be called from within an SQL query. Most commercial databases nowadays support similar extensions. The importance of sequence query processing was also discussed in [28, 29], where several ways to specify patterns as part of SQL, as well as their query optimization, are presented. Pattern queries have also been used for evaluating event streams using a NFA-based evaluation method [30]. However, all of the above works focus on patterns for time-series or event streams data, which are not explicitly designed for handling trajectories and spatio-temporal patterns.

In the moving object data domain, patterns have been examined in the context of query language and modeling issues [32, 34, 53, 54], as well as query evaluation algorithms [31, 33]. In [32], the authors propose the use of spatio-temporal patterns as a systematic and scalable query mechanism to model complex object behavior in space and time. The work in [53] presents a powerful query language able to model complex pattern queries using a combination of logical functions and quantifiers. Tao and Papadias [55] focuses on the *movement direction* of patterns, while queries related to the *relative movement* of objects are examined in the Relative Motion (REMO) system [56]. In [33], it is examined incremental ranking algorithms for simple spatio-temporal pattern queries. Those queries consist of range and *NN* predicates specified using only *fixed* regions. The FlexTrack system differs from [33] since it provides a more general and powerful query framework where queries can involve both fixed and *variable* regions, as well as regular expression structures (repetitions, negations, optional structures, etc) and explicit ordering of the predicates along the temporal axis. Moreover, the language used in the FlexTrack system introduces explicit ordering of the predicates along the temporal axis, which allows the users to specify ordering constraints like "immediately after" or "immediately before" between predicates. In [31], a *KMP*-based algorithm [57] is described to process motion patterns. This work, however, focuses only on range spatial predicates, and it cannot handle *explicit* and *implicit* temporal ordering of query predicates. Furthermore, the evaluation of pattern queries of this work is evaluated as a sequential scanning over the list of all trajectories stored in the database: each trajectory is checked individually, which becomes prohibitive for large trajectory databases.

While the use of variable predicates in specifying patterns greatly improves the expressive power, the query evaluation becomes more challenging. This is because variable spatio-temporal predicates provide many more opportunities for matching the pattern query to a specific trajectory in the repository, i.e., by simply changing the variable bindings in the pattern during the evaluation process. In the next sections we describe the FlexTrack system that addresses all of the above challenges.

2.3 The FlexTrack Pattern Query Language

We assume that a trajectory T_{id} of a moving object O_{id} is stored in the database as an ordered sequence of w pairs $\{(l_1, t_1), \ldots (l_w, t_w)\}$, where l_i is the moving object location recorded at timestamp t_i ($l_i \in \mathbb{R}^d$, $t_i \in \mathbb{N}$, $t_{i-1} < t_i$, and $0 < i \leq w$).

Figure 2.1 illustrates an example of trajectory for moving object mo_{id} with a sequence of 11 locations, from (t_1, t_1) to (t_{11}, t_{11}).

Such raw trajectory data is collected from applications (e.g., GPS, mobile devices) and stored in the database repository. Typically, monitored objects report their position to the data collection device using data packets containing a tuple $\langle T_{id}, l_i, t_i \rangle$. Depending on the application, objects may report continuously or simply when they change their location. For instance, if a trajectory is represented by a function f the (l_i, t_i) pairs can be created by sampling f at discrete timestamps [47].

We further assume that the spatial domain is partitioned to a fixed set Σ of non-overlapping regions. Figure 2.2 illustrates a hierarchical region alphabet with three layers, where the user has the ability to define queries with finer alphabet granularity (*zoom in*) for the portions of greater interest, and higher granularity (*zoom out*) elsewhere. Regions correspond to areas of interest (e.g., *school districts*, *airports*, *parks*, *city malls*) and form the alphabet used in our pattern query language. In the following we use capital letters to represent the region alphabet, $\Sigma = \{A, B, C, \ldots\}$.

A general pattern query $\mathcal{Q} = (\mathcal{S} [\bigcup \mathcal{D}])$ consists of a sequential pattern $\mathcal{S}$ and an optional set of constraints $\mathcal{D}$. The $\mathcal{S}$ component corresponds to a sequence of spatial-temporal predicates, and $\mathcal{D}$ represents a collection of distance functions (e.g., *NN*, *RNN* and their variations) or constrains (e.g., $@x = \{G, C, B\}$, $@y! = @z$) that may contain regions defined in $\mathcal{S}$. A trajectory matches the pattern query $\mathcal{Q}$ if it satisfies both $\mathcal{S}$ and $\mathcal{D}$. We first describe how a pattern $\mathcal{S}$ is formed and then elaborate on the distance constraints $\mathcal{D}$.

A pattern $\mathcal{S}$ is expressed as a path expression of an arbitrary number of spatio-temporal predicates $\mathcal{P}$, as defined in Fig. 2.3. The "!" defines the **negation** operator, "#" the **optional** modifier, "+" the **one or more repetition** modifier, "*" the **zero or more repetition** modifier, and "?" the **wild-card**. Within the pattern $\mathcal{S}$, the wild-card "?" is used to specify "don't care" parts in a trajectory's lifetime and can be of two types:

1. "$?^+$": one or more occurrences of any region predicate (e.g., $\mathcal{P}_i.?^+.\mathcal{P}_{i+1}$ implies that the predicate $\mathcal{P}_{i+1}$ is satisfied after predicate $\mathcal{P}_i$ with one or more regions visited between them), or;
2. "$?^*$": zero or more occurrences of any region visit (e.g., $\mathcal{P}_i.?^*.\mathcal{P}_{i+1}$ which implies that the predicate $\mathcal{P}_{i+1}$ can be satisfied any time after predicate $\mathcal{P}_i$).

The sequence of predicates in $\mathcal{S}$ is defined recursively by $\mathcal{S}.\mathcal{S}$, where the sequencer "." appears between every two consecutive spatio-temporal predicates $\mathcal{P}$ in $\mathcal{S}$. Each spatio-temporal predicate $\mathcal{P}_i \in \mathcal{S}$ is defined by a triplet $\mathcal{P}_i = \langle op_i, \mathcal{R}_i, [int_i] \rangle$. Here $\mathcal{R}_i$ corresponds to a predefined spatial region or a *variable*, i.e., $\mathcal{R}_i \in \Sigma \cup \Gamma$, where Γ is the set of variables (described below). The operator op_i describes the topological relationship that a trajectory T_{id} and the spatial region $\mathcal{R}_i$ must satisfy over the (optional) time interval int_i. In particular, we use the topological relationships described in [32]; examples of such operators are the relations *Equal*, *Inside*, *Touch*, *Meet*, among others. Given a trajectory T_{id} and a region $\mathcal{R}_i$, the operator op_i returns a boolean value $\mathbb{B} \equiv \{true, false\}$ whether the trajectory T_{id}

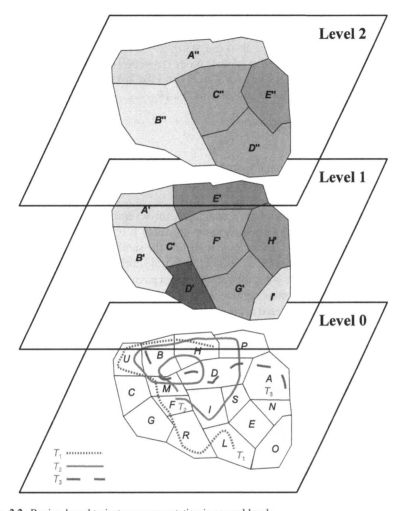

Fig. 2.2 Region-based trajectory representation in several levels

and the region $\mathcal{R}_i$ satisfy the topological relationship op_i (e.g., an *Inside* operator will be *true* if the trajectory was sometime inside region $\mathcal{R}_i$ during time interval int_i). For simplicity in the following we assume that the spatial operator is set to *Inside* and it is thus omitted from the query examples.

A predefined region $\mathcal{R}_i \in \Sigma$ is explicitly specified by the user in the query predicate (e.g., "the convention center"). In contrary, a *variable* denotes an arbitrary region and it is denoted by a lowercase letter preceded by the "@" symbol (e.g., @x). A variable region is defined using symbols in Γ, where $\Gamma = \{@a, @b, @c, \ldots\}$. Unless otherwise specified, a *variable* takes a single value (instance) from Σ (e.g., @a = C); however, in general, one can also specify the possible values of a *variable* as a subset of Σ (e.g., "any city district with museums"). Conceptually, *variables*

$$\mathcal{Q} \to (\mathcal{S} \, [\bigcup \mathcal{D}])$$
$$\mathcal{S} \to \mathcal{S}.\mathcal{S} \mid \mathcal{P} \mid !\mathcal{P} \mid \mathcal{P}^{\#} \mid ?^{+} \mid ?^{*}$$
$$\mathcal{P} \to \langle op, \mathcal{R} \, [, t] \rangle$$
$$op \to disjoint \mid meet \mid overlap \mid equal \mid inside \mid contains \mid covers \mid coveredBy$$
$$\mathcal{R} \in \{\Sigma \cup \Gamma\}, \, \Sigma = \{A, B, C, \ldots\}, \, \Gamma = \{@a, @b, @c, \ldots\}$$
$$t \to (t_{from} : t_{to}) \mid t_s \mid t_r$$

Fig. 2.3 The flexible pattern query language

work as placeholders for explicit spatial regions and can become instantiated (bound to a specific region) during the query evaluation in a process similar to unification in logical programming.

Moreover, the same *variable* $@x$ can appear in several different predicates of pattern $\mathcal{S}$, referencing to the same region everywhere it occurs. This is useful for specifying complex queries that involve revisiting the same region many times. For example, a pattern like $\mathcal{S} = \{@x.?^*.B.@x\}$ finds trajectories that started from some region (denoted by variable $@x$), then at some point passed by region B and immediately after they visited the same region they started from. Note that for our purposes, wild-card "?" is also considered a variable; however it refers to any region, and not necessarily the same region if it occurs multiple times within a pattern $\mathcal{S}$.

Finally, a predicate $\mathcal{P}_i$ may include an explicit temporal constraint int_i in the form of an interval, which implies that the spatial relationship op_i between a trajectory and region $\mathcal{R}_i$ should be satisfied in the specified time interval int_i (e.g., "passed by area B between 10 am and 11 am"). If the temporal constraint is missing, we assume that the spatial relationship can be satisfied any time in the duration of a trajectory lifespan. For simplicity we assume that if two predicates $\mathcal{P}_i, \mathcal{P}_j$ occur within pattern $\mathcal{S}$ (where $i < j$) and have temporal constraints int_i, int_j, then these intervals do not overlap and int_i occurs before int_j on the time domain.

However, spatio-temporal predicates cannot answer queries with constraints (e.g., "best-fit" type of queries—like *NN* and its related—that find trajectories which best match a specified pattern). This is because topological predicates are binary and thus cannot capture distance based properties of the trajectories. The optional $\mathcal{D}$ part of a general query $\mathcal{Q}$ is thus used to describe distance-based or other constraints among the *variables* used in the $\mathcal{S}$ part. A simple kind of constraint can involve comparisons among the used variables (e.g., $@x != @y$). More interesting is the distance-based constraint which has the form $(Aggr(d_1, d_2, \ldots); \theta)$ and is described below.

For simplicity in the following we assume Euclidean distance (L_2) but other distances, like *Manhattan* (L_1), *Infinity* (L_∞), among others, can also be used. Consider for example a $\mathcal{Q}$ query whose pattern $\mathcal{S}$ contains three *variables* $@x, @y, @z$, i.e., $\mathcal{S} = \{A.?^*.B.@x.@y.C.?^*.@z\}$. Among the trajectories that satisfy $\mathcal{S}$, the user may specify that in addition, the sum of the distance between regions $@x$ and $@y$ and the distance between $@z$ and a fixed region E is less than 100 feet. Hence $\mathcal{D}$ contains a collection of distance terms $d_1, d_2, \ldots, d_n$ where term d_i represents the distance between two *variable* regions or between a *variable* region and a fixed one. In our example there are two distance terms: $d_1 = d(@x, @y)$ and $d_2 = d(@z, E)$.

Distance terms need to be aggregated into a single numerical value using an aggregation function (depicted as $Aggr$ in the formal definition of $\mathcal{D}$). In the previous example $Aggr = Sum$, but other aggregators, like Avg, Min, Max, etc., can also be used. The aggregated numerical score for each trajectory still needs to be mapped to a binary value so as to determine if the trajectory satisfies $\mathcal{D}$. This is done by the θ operator defined in $\mathcal{D}$. This operator can be a simple check function (using =, $\leq$ and others). In our example, θ corresponds to "<100 feet" and returns *true* for all trajectories whose aggregate distance is less than 100 feet. It is also possible to use other θ operators, e.g., Min, Max, Top-k, etc. In the previous example, if the θ operator is changed to Top-k, the query will return *true* only for the trajectories with the Top-k aggregated distances. For simplicity of the description, in the remainder of this work we use $Aggr = Sum$ and $\theta = Min$ (which corresponds to a *NN* query).

The use of *variables* in describing both the topological predicates and the numerical conditions provides a very powerful language to query trajectories. To describe a query, the user can use fixed regions for the portions of the trajectory where the behavior should satisfy known (strict) requirements. The user can also employ *variables* for portions where the exact behavior is not known, but can be described by a sequence of *variables* and the constraints between them. The ability to use the same *variable* many times in the query allows for revisiting areas, while the ability to refer to these *variables* in the distance functions allows for easy description of *NN* and related queries. It is exactly this *flexibility* allowed by the use of *variables* in selecting trajectories that led to the term *flexible pattern queries*.

2.4 The FlexTrack Query Evaluation System

To simplify the presentation we first start with the evaluation of the spatial predicates for a pattern $\mathcal{S}$. Later we extend the discussion to cover queries that in addition contain distance constraints $\mathcal{D}$. Finally we present the incorporation of time constraints inside the pattern query $\mathcal{Q}$.

For simplicity we assume that the space is partitioned into 2-dimensional non-overlapping regions, as exemplified in Fig. 2.4a. To efficiently evaluate flexible pattern queries we will facilitate two lightweight index structures in the form of ordered lists (Fig. 2.4b, c), which are stored in addition to the raw trajectory data (Fig. 2.4d). There is one *region-list* per region and one *trajectory-list* per trajectory. The *region-list* $\mathcal{L}_A$ of a given region A acts as an inverted index that contains all trajectories that passed by region A. Each entry in $\mathcal{L}_A$ is a record that contains a trajectory identifier T_{id}, the time interval (*t-entry*: *t-exit*) during which the moving object was inside A, and a pointer to the *trajectory-list* of T_{id}. If a trajectory visits a given region A multiple times in different time intervals, we store a record for each visit. Records in a *region-list* are ordered first by the trajectory-id T_{id} and then by *t-entry*.

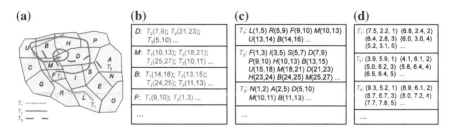

Fig. 2.4 Example of a region-based trajectory representation with three trajectories $\{T_1, T_2, T_3\}$.
(**a**) Partitioned spatial domain, (**b**) Region-list, (**c**) Trajectory-list, (**d**) Raw trajectory archive

For example, in Fig. 2.4 the *region-list* entry for the region D is $\{T_2(7, 9); T_2(21, 23); T_3(5, 10); \ldots\}$.

In order to fast prune trajectories that do not satisfy the pattern S, each trajectory is approximated by the sequence of regions it visited. A record in the *trajectory-list* of trajectory T_{id} contains the region and the time interval (*t-entry:t-exit*) during which this region was visited by T_i, ordered by *t-entry*. In Fig. 2.4c the *trajectory-list* entry for T_2 is $\{F(1, 3); I(3, 5); S(5, 7); D(7, 9); P(9, 10); H(10, 13); B(13, 15); U(15, 18); M(18, 21); D(21, 23); H(23, 24); B(24, 25); M(25, 27)\}$. Note that records from a *region-list* index point to the corresponding records in a *trajectory-list* index. For example, the record $T_2(21, 23)$ in the *region-list* $\mathcal{L}_D$ contains a pointer to the page in the *trajectory-list* of T_2 that contains the corresponding record $D(21, 23)$.

Since *variables* in pattern S can take values from the whole set Σ of regions, we need a representation of each trajectory using the alphabet elements in Σ. While one could always use the raw trajectory data, it is more efficient to maintain a region representation of each trajectory to fast prune trajectories that do not satisfy the pattern S. That is, each trajectory is approximated by the sequence of regions it visited. This compact representation of each trajectory is stored in the *trajectory-list* index. A record in the *trajectory-list* of trajectory T_{id} contains the region and the time interval (*t-entry:t-exit*) during which this region was visited by T_i, ordered by *t-entry*. Figure 2.4b, c depict various *region–lists* and *trajectory-lists*. Note that records from a *region-list* index point to the corresponding records in a *trajectory-list* index. For example, consider the *region-list* $\mathcal{L}_D$ of region D and a record in this list for trajectory T_2 with interval (t_1, t_2). The pointer included in this record points to the page in the *trajectory-list* of T_2 that contains the corresponding record $D(t_1, t_2)$.

The only requirement for the region partitioning is that regions should be non-overlapping. In practice, there may be a difference between the regions presented to the user and what lists are created. In such scenarios we use uniform grid and overestimate a region by approximating it with the smallest collection of grid cells which completely encloses it. False positives may be generated from regions that do not completely fit the set of covering grid cells; however, they can be removed with a verification step using the original trajectory data. Finding the best grid granularity can be done by an optimization process which combines the number of grid cells and the total overestimated area into a single objective function. Moreover, instead of a uniform grid, one could facilitate instead a dynamic space partitioning structure (for instance, adaptive grid files [58], kdb-trees [59], BANG files [60], among many other index structures) that assigns grid cells sizes according to the data density. Then, dense areas will have finer cells which in return allow for better approximation of the regions, and thus fewer false positives are generated.

For evaluating pattern queries we propose two different strategies: (1) the *Index Join Pattern (IJP)* is based on a merge-join operation performed over the *region-lists* corresponding to every fixed predicate in the pattern S; (2) the *Dynamic Programming Pattern (DPP)* performs subsequence matching between the pattern S and the trajectory approximations stored as the *trajectory-lists*. Both algorithms use the same two indexing structures for pruning purposes, but in different ways: *IJP* uses the *region-lists* for pruning and the *trajectory-lists* for the variable binding; *DPP* uses mainly the *trajectory-lists* for the subsequence matching and performs an intersection-based pruning on the *region-lists*. Which algorithm would behave better will thus depend on the pruning capabilities provided by its main index; this in turn depends on the trajectory archive and the query characteristics.

2.4.1 The Index-Join Pattern Algorithm (IJP)

2.4.1.1 Spatial Predicate Evaluation

We start with the case where the pattern S does not contain any explicit temporal constraints. In this scenario, the pattern S specifies the order by which its predicates, whether fixed or variable, need to be satisfied. Assume S contains m predicates and let S_f denote the set of n fixed predicates, while S_v denotes the set of r variable predicates ($m = n + r$). The evaluation of S with the *IJP* Algorithm can be divided in two steps: (**i**) the algorithm evaluates the set S_f using the *region-list* index to fast prune trajectories that do not qualify for the answer; (**ii**) then the collection of *candidate* trajectories is further refined by evaluating the set of S_v.

(i) **Fixed predicate evaluation:** All n fixed predicates in S_f can be evaluated *concurrently* using an operation similar to a merge-join among their *region-lists* $\mathcal{L}_i, i \in 1 \dots n$. Records from these n lists are retrieved in sorted order of T_{id} and then joined by their T_{id}'s. Records are pruned using the trajectory *ids* and the temporal intervals (*t-entry*: *t-exit*). In each list $\mathcal{L}_i$ we keep a pointer p_i that points to the record currently considered for the join. This pointer scans the list $\mathcal{L}_i$ starting from the top.

If the same region appears more than once in the pattern S, a duplicate pointer traversing that *region-list* is used for each region appearance in S. For example, to process the pattern $S = \{?^*.M.D.M\}$, the *region-lists* of M and D are accessed using one pointer for *region-list* $\mathcal{L}_D$ (p_D) and two pointers for traversing *region-list* $\mathcal{L}_M$ (p_{M_1} and p_{M_2}). If a trajectory-id T_{id} appears in *all* of the n *region-lists* involved in the pattern query, and their corresponding time intervals in all n *region-lists* satisfy the ordering of the predicates in S, this T_{id} is saved as a candidate solution. Algorithm 1 shows the IJP algorithm.

There are cases during the merge-join operation where records from a *region-list* can be skipped, thus resulting in faster processing. For example, assume that predicate $P_i \in S$ (corresponding to the *region-list* $\mathcal{L}_i$) is before predicate $P_j \in S$ (corresponding to $\mathcal{L}_j$). Further assume that in list $\mathcal{L}_i$ the current record considered for the join has trajectory identifier T_r, while in list $\mathcal{L}_j$ the current record considered has trajectory identifier T_s. If $T_s < T_r$, processing in list $\mathcal{L}_j$ can skip all its records with $T_{id} < T_r$. That is, the pointer p_j in list $\mathcal{L}_j$ can advance to the first record with $T_{id} \geq T_r$. Essentially, predicate $\mathcal{P}_i$ cannot be satisfied by any of the trajectories in $\mathcal{L}_j$ with smaller T_{id} than T_r. Since records in a *region-list* are sorted by T_{id}, $\mathcal{L}_i$ does not contain trajectories with smaller identifiers than r.

Similarly, when a record from the same trajectory, e.g., T_s is found in two *region-lists*, e.g., $\mathcal{L}_i$, $\mathcal{L}_j$, the algorithm checks whether the corresponding time intervals of the records match the order of predicates in the pattern S. Hence, a trajectory that satisfies S must visit the region of $\mathcal{L}_i$ before visiting the region of $\mathcal{L}_j$. If the record of T_s in $\mathcal{L}_i$ has *t-entry* that falls after the corresponding *t-entry* of T_s in list $\mathcal{L}_j$, then this record can be skipped in $\mathcal{L}_i$ since it cannot satisfy the pattern query. Since *region-lists* are stored in ordered way, advancing a *region-list* forward to a specific location stamp by T_{id} or by $(T_{id}, t\text{-}entry)$ can be easily implemented using an index B^+-tree on the $(T_{id}, t\text{-}entry)$ composite attribute.

Example: The first step of *IJP* algorithm is illustrated using the example in Fig. 2.5. Assume the pattern S in the query Q contains three fixed (M, D, M) and three *variable* predicates $(?^+, @x, @x)$, as in:

$$S = \{?^+ . @x . ?^* . M . ?^* . D . ?^* . @x . ?^* . M\}$$

This pattern looks for trajectories that first visited an arbitrary region (denoted by $?^+$) one or more times, then visited some region denoted by *variable* $@x$, then (after visiting zero or more regions) it visited region M, then region D and then visited again the same region $@x$ before finally returning to M. The first step of the merge-join algorithm uses the *region-list* for M and D ($\mathcal{L}_M$ and $\mathcal{L}_D$). For simplicity, instead of using two separate pointers in list $\mathcal{L}_M$, Fig. 2.5 depicts two copies of list $\mathcal{L}_M$, namely $\mathcal{L}_{M_1}$ and $\mathcal{L}_{M_2}$. Conceptually, $\mathcal{L}_{M_1}$ represents the first occurrence of M in S before D, and $\mathcal{L}_{M_2}$ the second occurrence of M after D.

The algorithm starts from the first record in list $\mathcal{L}_{M_1}$, namely $T_1(10, 13)$. It then checks the first record in list $\mathcal{L}_D$, i.e., trajectory T_2. We can deduce immediately that T_1 is not a candidate trajectory, since it does not appear in the list of $\mathcal{L}_D$, so we can skip T_1 from the $\mathcal{L}_{M_1}$ list and continue with the next record there, trajectory $T_2(18, 21)$. Since $T_2(7, 9)$ in list $\mathcal{L}_D$ has interval before $(18, 21)$, list $\mathcal{L}_D$ moves to its next record $T_2(21{:}23)$. These two occurrences of T_2 coincide with the pattern $M . ?^* . D$ of S, so we need to check if T_2 passes again by region M. Thus we consider the first record of list $\mathcal{L}_{M_2}$, namely trajectory $T_1(10{:}13)$. Since it is not from T_2 it cannot be an answer so list $\mathcal{L}_{M_2}$ advances to the next record $T_2(18, 21)$. Now pointers in all lists point to records of T_2. However, $T_2(18, 21)$ in $\mathcal{L}_{M_2}$ does not satisfy the pattern since its time interval should follow the interval $(21,23)$ of T_2 in D. Hence $\mathcal{L}_{M_2}$ is advanced to the next record, which happens to be $T_2(25,27)$. Again we have a record from the same trajectory T_2 in all lists and this occurrence of T_2 satisfies the temporal constraints and thus the pattern $\mathcal{S}$. As a result, trajectory T_2 is kept as a candidate in U. The processing moves to the next record in list $\mathcal{L}_{M_1}$, namely $T_2(25,27)$. However, this record cannot satisfy the pattern $\mathcal{S}$ so it is skipped. Eventually $\mathcal{L}_{M_1}$ will consider $T_3(10,11)$ which causes list $\mathcal{L}_D$ to move to $T_3(5,10)$. Trajectory T_3 cannot satisfy the temporal constraint, so it is skipped from list $\mathcal{L}_D$ and the algorithm terminates since one of the lists reached its end.   □

In the case where the region partitioning is represented internally by a grid of smaller cells, Algorithm 1 can still be applied. But to evaluate such region's predicate, we need first to *materialize* a sorted list from all *cell-lists* involved in this region. However, since the individual *cell-lists* participating in the enclosure are already ordered by trajectory-id T_{id}, the sort order can be materialized *on the fly* by feeding

---

Algorithm 1 *IJP*: Fixed Spatial Predicates

Input: Pattern $\mathcal{S}$
Output: Trajectories satisfying $\mathcal{S}_f$
1: $n \leftarrow |\mathcal{S}_f|$
2: **for** $i \leftarrow 1$ to n **do**
3:    Initialize $\mathcal{L}_i$ with the *cell-list* of $\mathcal{P}_i$
4: Candidate Set $U \leftarrow \emptyset$
5: **for** $w \leftarrow 1$ to $|\mathcal{L}_1|$ **do**
6:    $p_1 = w$
7:    **for** $j \leftarrow 2$ to n **do**
8:       **if** $\mathcal{L}_1[w].id \notin \mathcal{L}_j$ **then**
9:          **break**
10:      Let k be the first entry for $\mathcal{L}_1[w].id$ in $\mathcal{L}_j$
11:      **while** $\mathcal{L}_1[w].id = \mathcal{L}_j[k].id$ **and** $\mathcal{L}_{j-1}[p_{j-1}].t > \mathcal{L}_j[k].t$ **do**
12:         $k \leftarrow k+1$
13:      **if** $\mathcal{L}_1[w].id \neq \mathcal{L}_j[k].id$ **then**
14:         **break**
15:      **else** $p_j = k$
16:    **if** $\mathcal{L}_1[w]$ qualifies **then**
17:      $U \leftarrow U \cup \mathcal{L}_1[w].id$

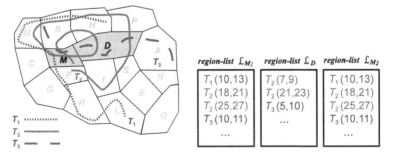

Fig. 2.5 Trajectory examples T_1, T_2 and T_3 satisfying two spatial fixed predicates: M and D

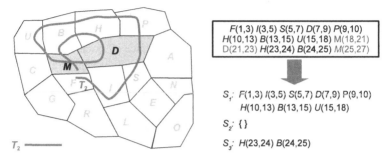

Fig. 2.6 Example on the *IJP* segmentation step for T_2 ($S_2 = \emptyset$)

the algorithm with the record that has the smallest T_{id} among the heads of the participating *cell-lists*. Hence the algorithm proceeds without having to actually sort the participating *region-lists*.

(ii) **Variable predicate evaluation:** The second step of the *IJP* algorithm evaluates the *variable* predicates r in $\mathcal{S}_v$ over the set of candidate trajectories U generated in the first step. For a fixed predicate its corresponding *region-list* contains all trajectories that satisfy it. However, *variable* predicates can be bound to any region, so one would have to look at all *region-lists*, which is not realistic. We will again need one list per each *variable* predicate (termed *variable-list*), however such *variable-lists* are not precomputed like the *region-lists*. Rather, they are created *on the fly* using the candidate trajectories filtered from the fixed predicate evaluation step.

To populate a *variable-list* for a *variable* predicate $P_j \in \mathcal{S}_v$, we compute the possible assignments for *variable* P_j by analyzing the *trajectory-list* for each candidate trajectory. In particular, we use the time intervals in a candidate trajectory to identify which portions of the trajectory can be assigned to this particular *variable* predicate. An example is shown in Fig. 2.6, using the candidate trajectory T_2 from Fig. 2.5. From the previous step we know that T_2 satisfies the fixed predicates at the following regions: $M(18, 21)$, $D(21, 23)$, $M(25, 27)$ (shown in bold in the *trajectory-list* of T_2). Using the pointers from the *region-lists* of the previous step, we know where the matching regions are in the *trajectory-list* of T_2. As a result, T_2 can be conceptually partitioned is three segments $\{S_1, S_2, S_3\}$, as shown in Fig. 2.6. Note that S_2 is empty since there is no region between $M(18, 21)$ and $D(21, 23)$.

These trajectory segments are used to create the *variable-lists* by identifying the possible assignments for every *variable*. Since a *variable*'s assignments need to maintain the pattern, each *variable* is restricted by the two fixed predicates that appear before and after the *variable* in the pattern. All *variables* between two fixed predicates are first grouped together. Then for every group of *variables* the corresponding trajectory segment (the segment between the fixed predicates) is used to generate the *variable-lists* for this group. Grouping is advantageous, since it can create *variable* lists for multiple *variables* through the same pass over the trajectory segments. Moreover, it ensures that the *variables* in the group maintain their order consistent with the pattern $\mathcal{S}$.

Assume that a group of *variable* predicates has w members. Each trajectory segment that affects the *variables* of this group is then "streamed" through a window of size w. The first w elements of the trajectory segment are placed in the corresponding predicate lists for the *variables*. The first element in the segment is then removed and the window shifts by one position. This proceeds until the end of the segment is reached. In the above example, there are two groups of *variables*: the first consists of *variables* $?^+$ and $@x$ in that order (i.e., $w = 2$), while the second group has a single member $@x$ ($w = 1$). Figure 2.7 depicts the first three steps in the *variable* list generation for the group of *variables* $?^+$ and $@x$. This group streams through segment S_1, since it is restricted on the right by the fixed predicate M in pattern S. Each list is shown under the appropriate *variable*. A different *variable* list will be created for the second group with *variable* $@x$, since this group streams through segment S_3 (the second $@x$ *variable* is restricted by fixed predicates D and M).

The generated *variable-lists* are then joined in a way similar to the previous step of fixed predicates. Because the *variable-lists* are populated by trajectory segments coming from the same trajectory (in our example trajectory T_2), the join criteria checks only if the ordering of pattern S is satisfied. In addition, if the pattern contains *variables* with the same name (e.g., $@x$), the join condition verifies that they are matched to the same region and time interval.

Complexity Analysis for variable predicate evaluation: Assume that the fixed predicate evaluation step generates k candidate trajectories in U and let l denote the maximum trajectory segment length. The worst case scenario is when all *variable*

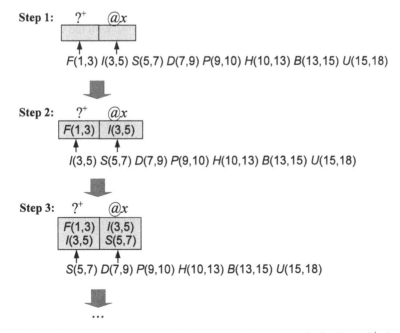

Fig. 2.7 Example with three steps on how the *IJP* creates the *variable* list for T_2 and $?^+.@x$

lists have length l. Thus the *variable* predicate evaluation in the worst case scenario is $O(klr)$.

Explicit Temporal Constraints: The *IJP* algorithm can easily support explicit temporal constraints (assigned to the spatial predicates) by incorporating them as extra conditions in the join evaluations among the list records.

2.4.1.2 Distance-Based Constraints Evaluation

The evaluation of distance constraints $\mathcal{D}$ in a pattern query $\mathcal{Q}$ is performed as a post filtering step after evaluating pattern $\mathcal{S}$. The intuition is that the spatial predicates in $\mathcal{S}$ will greatly reduce the number of candidate trajectories that need to be examined by the distance-based algorithm. Nevertheless, since the distance terms contain *variables*, there are still many possibilities to bind the values of these *variables*. The *IJP* algorithm has the advantage of re-using the *variable* lists created during the spatial predicate search. These lists effectively enumerate all possible value bindings. However, instead of using a *brute force* approach that will examine all possible bindings, the *IJP* approach uses a variation of the *Threshold Algorithm* (TA) [61] to examine these possibilities in an incremental ordered fashion. As a result, it avoids examining all possible bindings.

Regarding the *IJP* approach, assume that the $\mathcal{S}$ evaluation has returned a collection of trajectories $\mathcal{T}$. For each *variable* in $\mathcal{S}$ one *variable-list* per trajectory in $\mathcal{T}$ is also created. All *variable-lists* for a given *variable* are concatenated and sorted, first by region *id* and then by trajectory *id*. Note that the same region may be associated with different trajectory *ids*. For simplicity, consider the scenario where the distance terms are combinations of a *variable* with a fixed region (i.e., $d(@x, A)$). The case where the distance term contains two *variables* is omitted for brevity.

For each distance term in $\mathcal{D}$ a separate list is created. As with the *variable-lists*, *distance-lists* are also computed *on the fly*. The idea is to incrementally examine the vicinity around the fixed *variable* of each distance term d_i. To evaluate distances between regions, we use the uniform grid that has been introduced in Sect. 2.4. We will use the distance between grid cells to lower bound the Euclidean distance between regions.

For example, given a term $d(@x, A)$, in the first iteration we examine the grid cells, and the regions approximated with those grid cells, that are one cell away from the grid approximation of region A. The next iteration will expand the vicinity by one cell, and so on. When we discover a region which appears also in the sorted concatenated list for $@x$, we load all the corresponding trajectory *ids* and place them in the list for this distance term. As the lists for all distance terms in $\mathcal{D}$ have been created incrementally, the *TA* algorithm finds the trajectory that appears in all *distance-lists* and minimizes the sum of the distances.

2.4.2 The Dynamic Programming Pattern Algorithm (DPP)

We now proceed with the description of the DPP Algorithm. The *DPP* algorithm is divided into two steps: (i) *Trajectory Selection* and (ii) *Matching*. The first step use the *trajectory-lists* to select a candidate set of trajectories $\bar{T}$ based on the fixed predicates in pattern S. The second step uses pattern matching to prune trajectories that do not match the correct sequence order of predicates in S. It also checks for appropriate *variable* bindings with possible verification on duplicate *variables* in S. The pseudo code for the *DPP* algorithm is described in Algorithm 2.

(i) **Trajectory Selection:** For each *region-list* of a fixed region that appears in S, we select the *ids* T_{id} for all trajectories that visited this region. Candidate set $\bar{T}$ is computed by intersecting the collected set of T_{id} (one set per unique region). That is, $\bar{T}$ contains *ids* of the trajectories that have visited (independently of what order) all the regions in S. Nevertheless, since no order of these appearances has been verified, a further verification step must be performed on each $T' \in \bar{T}$ to enforce the *order* of S. This verification step is performed using a dynamic programming approach.

(ii) **Matching:** For each trajectory $T' \in \bar{T}$ a dynamic programming matrix $\mathcal{M}$ (function *BuildDPM*) is first created; it will later retrieve the matches of S in the trajectory T' (function *ScanDPM*). The $\mathcal{M}$ matrix enables the *DPP* algorithm to match **all** occurrences of the pattern S in T' in the specified order defined in S. Matrix $\mathcal{M}$ has a column j for each region visited by the trajectory T'. Multiple visits to the same region are represented with multiple columns in $\mathcal{M}$, as it is stored the same way in the *trajectory-list* index. The rows i in the matrix correspond to the predicates $P_i \in S$. Therefore, the size of $\mathcal{M}$ is $|S|.|T'|$. The value in each cell in $\mathcal{M}[i][j]$ is computed based on the predicate P_i and the j-th element in the region approximation of the trajectory T' denoted as T'_j (this is the j-th element in the *trajectory-list* of T').

It should be noted that when the pattern S contains only fixed spatial predicates, the matrix $\mathcal{M}$ can be reduced by eliminating the regions in T' that are not present in S. This optimization does not compromise the sequence of patterns found, because for each R_j in T' the attribute $(t - entry_j : t - exit_j)$ is also kept.

Each matrix cell can a take numerical value in the range $(-|S|; |S|)$. The absolute value stored in the matrix entries corresponds to the length of the longest match between the pattern S and the trajectory approximation T' discovered so far. A negative number in $\mathcal{M}[i][j]$ denotes a match between the pattern P_i and the trajectory region R_i, and its absolute value is the length of the longest match found so far. In this way, the matrix $\mathcal{M}$ is used to store both the match occurrences, represented with negative value, and the length of each match, the absolute values in $\mathcal{M}[i][j]$.

The matrix $\mathcal{M}$ is computed row by row, column by column, starting from the $\mathcal{M}[0][0]$ cell until the $\mathcal{M}[|S|][|T'|]$ cell. At every step the *BuildDPM* function compares the values of the current predicate P_i and the current region from the trajectory approximation T_j (the same as the T'_j). If there is no match between P_i and T_j, then the value of $\mathcal{M}[i][j]$ is the biggest absolute value among the neighbors ($\mathcal{M}[i-1][j]$ or $\mathcal{M}[i][j-1]$). If there is a match between P_i and T_j then the cell $\mathcal{M}[i][j]$ takes the value $|\mathcal{M}[i-1][j-1]| + 1$, but it is stored as a negative number indicating that the current pair P_i, T_j participates in the match.

The above description applies only for fixed spatial predicates. As for wild-card ($?^+, ?^*$) and variable ($@$) spatial predicates, the computation of the $\mathcal{M}[i][j]$ cell is done differently. Because such variables can be *bound* with any value of T_j, the value of $\mathcal{M}[i][j]$ is always computed as a "match". Therefore, the cell value is $-(|\mathcal{M}[i-1][j-1]|+1)$, as previously described. This phase does not handle the case where a pattern $\mathcal{S}$ contains *variables* that appear multiple times. This verification step is performed in the *ScanDPM* function. Instances of the same variable are "linked" in a backward way using a "pointer" (*link*) with the following constraint: $P_i.link \leftarrow P_j$ if $P_i = P_j$ and $i < j$. Because matrix $\mathcal{M}$ is verified for matching in a "backward" way (from $\mathcal{M}[|\mathcal{S}|][|T'|]$ to $\mathcal{M}[1][1]$ cell), the pointers are associated to the next occurrence in the pattern $\mathcal{S}$.

There is also a special case where the predicate P_i is *optional* in the pattern $\mathcal{S}$. In this case, the computation and further verification of matrix $\mathcal{M}$ has to consider the case where P_i does not match T_j. To deal with this, another attribute $P_i.idx$ is associated with each predicate in $\mathcal{S}$. Basically, this attribute stores the position of each predicate P_i in cases the optional predicate does not match with any T_j. This idx attribute is defined in the following manner:

$$P_i.idx \leftarrow \begin{cases} 1 & \textbf{if } i = 1 \\ P_{i-1}.idx & \textbf{if } P_i.type = \{?^*, ?^\#\} \\ P_{i-1}.idx + 1 & \textbf{otherwise} \end{cases}$$

Algorithm 2 *DPP*: Fixed and Variable Spatial Predicates

Input: Pattern $\mathcal{S}$ that consists of predicates $\mathcal{P}_i$
Output: Trajectories satisfying $\mathcal{S}$
1: Let $\bar{T}$ be the set of candidate trajectories from *trajectory-list* having all fixed predicates in $\mathcal{S}$
2: Answer Set $\mathcal{A} \leftarrow \emptyset$
3: **for each** trajectory $T' \in |\bar{T}|$ **do**
4:     *BuildDPM*$(T', \mathcal{S})$
5:     **if** $Abs(\mathcal{M}[|\mathcal{S}|][|T'|]) \geq \mathcal{P}_{|\mathcal{S}|}.idx$ **then**
6:         *ScanDPM*$(|\mathcal{S}|, |T'|)$

Function: *BuildDPM*$(T, \mathcal{S})$
1: **for** $i \leftarrow 0$ to $|\mathcal{S}|$ **do**
2:     **for** $j \leftarrow 0$ to $|T|$ **do**
3:         **if** $i = 0$ **or** $j = 0$ **then** $\mathcal{M}[i][j] \leftarrow 0$
4:         **else**
5:             **if** $P_i.type$ is a Fixed Spatial Predicate **then**
6:                 **if** $P_i.R = T.R_j$ **then**
7:                     $\mathcal{M}[i][j] \leftarrow (-(Abs(\mathcal{M}[i-1][j-1])+1))$
8:                 **else**
9:                     $\mathcal{M}[i][j] \leftarrow Max(Abs(\mathcal{M}[i-1][j]), Abs(\mathcal{M}[i][j-1]))$
10:            **else**
11:                **if** $P_i.type = \{?^+, @\}$ **then**
12:                    $\mathcal{M}[i][j] \leftarrow (-(Abs(\mathcal{M}[i-1][j-1])+1))$
13:                **else**
14:                    **if** $i = P_i.idx$ **then**
15:                        $\mathcal{M}[i][j] \leftarrow Abs(\mathcal{M}[i-1][j])$
16:                    **else** $\mathcal{M}[i][j] \leftarrow (-(Abs(\mathcal{M}[i-1][j-1])+1))$

Function: *ScanDPM*(i, j)
1: **if** $i > 0$ **then**
2:     **for** $k \leftarrow j$ to $k \geq P_i.idx$ **downto** 1 **do**
3:         **if** $Abs(\mathcal{M}[i][k]) \geq P_i.idx$ **then**
4:             **if** $\mathcal{M}[i][k] \leq 0$ **then**
5:                 **if** $P_i.type = \{@\}$ **and** $Match[P_i.link] \neq T'.R_k$ **then continue**
6:                 $Match[i] \leftarrow T'.R_k$
7:                 **if** $P_{i-1}.type = \{?^*\}$ **then**
8:                     $ScanDPM(i - 1, k)$
9:                 **else**
10:                     $ScanDPM(i - 1, k - 1)$
11: **else** $\mathcal{A} \leftarrow \mathcal{A} \cup T'.id$

After the matrix $\mathcal{M}$ is fully computed, all the matches have to be searched. This is performed by the *ScanDPM* function that "searches" for negative numbers stored in $\mathcal{M}$; such numbers denote the occurrence of a match. The operation goes row by row, column by column in a direction opposite to the direction of construction, starting with the bottom right cell. If the last matrix cell $\mathcal{M}[i][|T'|]$ has an absolute value greater than the last idx in $\mathcal{P}$ (i.e., $Abs(\mathcal{M}[|\mathcal{S}|][|T'|]) \geq P_{|\mathcal{S}|}.idx$), then there is at least one match between $\mathcal{S}$ and T'. Otherwise we can safely prune the trajectory avoiding further processing. Because we are only interested in finding the longest and complete match between $\mathcal{S}$ and T', we only look for entries that have values greater or equal than the $S_i.idx$ index (smaller values indicate that there is a partial match but not a complete one). If the cell value is less than the current pattern index $S_i.idx$, then the function *ScanDPM* aborts the processing of the current row i.

If there is a match in $\mathcal{M}[i][j]$, then the function *ScanDPM* is called recursively to process the sub-matrix with bottom right corner $\mathcal{M}[i - 1][j - 1]$. If the predicate P_i is optional ($^\#$ and *) then the function is called for the $\mathcal{M}[i - 1][j]$ cell instead. The algorithm stops when all predicates in $\mathcal{S}$ are processed ($i = 0$), thus finding *all* possible matches of $\mathcal{S}$ in T'.

Complexity Analysis: The *BuildDPM* function calculates the value for each matrix cell just once. Let s denote the length of a trajectory T' in terms of number of regions visited. Then the matrix $\mathcal{M}$ has m rows ($|\mathcal{S}|$) and s columns, and the complexity of this method is $O(sm)$. The complexity of *ScanDPM* is $O(m + s)$ because at each step we move one step left-up diagonally or up (e.g., at least one of i and j is decremented). Therefore, the time complexity for processing a single trajectory T' with the *DPP* algorithm is $O(vsm)$, where $v = |\bar{T}|$ (i.e., the number of candidate trajectories produced from the ***trajectory selection*** step). The reader should note that the two algorithms produce candidate trajectory sets using different methods (*IJP* considers the temporal order and *DPP* does not); hence in the complexity analysis they are represented as k and v.

Explicit Temporal Constraints: When the pattern $\mathcal{S}$ has explicit temporal constrains int_i in its definition, the *DPP* algorithm only performs a check along with the match checks in order to satisfies int_i too (not shown in Algorithm 2). If only one of the above conditions is satisfied, then the value of $\mathcal{M}[i][j]$ is computed as not a match. Otherwise, it is computed as a match.

Example: We use the same example of pattern S in Fig. 2.5 to illustrate how the *DPP* algorithm works. Using the *region-list* the trajectory identifiers that have all the grids M and D are in $\bar{T} = \{T_2, T_3\}$. For each trajectory T' in $\bar{T}$, the matrix $\mathcal{M}$ is computed using the function *BuildDPM*. The computation of matrix $\mathcal{M}$ for T_2 and S appears in Table 2.1. Since $P_{|S|}.idx$ is 6, the *ScanDPM* function looks for cell values equal to $\mathcal{M}[10][j] \geq |-6|$ in the 10-th row of matrix $\mathcal{M}$. In *ScanDPM*, the $\mathcal{M}[10][13]$ cell passes the checks of the algorithm and the $\mathcal{M}[10][13]$ cell is stored as a match in *Match*[10] (M was found in the 13*th* column of T_2) and then the function *ScanDPM* is called for the $\mathcal{M}[9][12]$ matrix. Again, $\mathcal{M}[9][12]$ cell passes all the checks and it is called for $\mathcal{M}[8][12]$. Because P_8 is a *variable* (i.e., *variable* @x) and it is the first *variable* encountered so far, it passes the bounded value check (*link* test) and then it is bounded to the grid B. Then the function *ScanDPM* is called in the following sequence for entries in $\mathcal{M}$: $\mathcal{M}[7][11]$, $\mathcal{M}[6][10]$, $\mathcal{M}[5][10]$, $\mathcal{M}[4][9]$, $\mathcal{M}[3][8]$ and then for $\mathcal{M}[2][8]$, but it fails for this last one because the *link* test does not pass ($\mathcal{M}[2][8] \neq \mathcal{M}[8][12]$). Then it is called for $\mathcal{M}[2][7]$, and the *link* test satisfies because variable @x is bounded to grid B ($\mathcal{M}[2][7] = \mathcal{M}[8][12]$). Then *ScanDPM* is called for $\mathcal{M}[1][6]$ until j is 0. In the end, the pattern $?^+.B.?^*.M.?^*.D.?^*.B.?^*.M$ is found and added to $\mathcal{A}$. The backtracking also evaluates the $\mathcal{M}[8][11]$ cell and finds pattern $?^+.H.?^*.M.?^*.D.?^*.H.?^*.M$. Other calls for other entries are called, e.g., $\mathcal{M}[10][9]$ (-8), but they all fails to bound to other predicates in S. The 2 patterns found for the pattern S in trajectory T_2 are highlighted in Table 2.1 (symbol * for the first pattern found, symbol $^\circ$ for the second, and symbol $^\circledast$ when the entries are found for both of patterns).   □

Table 2.1 Matrix $\mathcal{M}$ for trajectory T_2 and pattern S

			T_2	F	I	S	D	P	H	B	U	M	D	H	B	M
			j	1	2	3	4	5	6	7	8	9	10	11	12	13
S	i	idx	0	0	0	0	0	0	0	0	0	0	0	0	0	0
$?^+$	1	1	0	$-1^\circledast$	$-1^\circledast$	$-1^\circledast$	$-1^\circledast$	$-1^\circledast$	-1^*	-1	-1	-1	-1	-1	-1	-1
@x	2	2	0	0	-2	-2	-2	-2	-2°	-2^*	-2	-2	-2	-2	-2	-2
$?^*$	3	2	0	0	2	-3	-3	-3	-3	-3°	$-3^\circledast$	-3	-3	-3	-3	-3
M	4	3	0	0	0	3	3	3	3	3	3	$-4^\circledast$	4	4	4	-4
$?^*$	5	3	0	0	0	3	-4	-4	-4	-4	-4	-4	-5	-5	-5	-5
D	6	4	0	0	0	0	-4	4	4	4	4	4	$-5^\circledast$	5	5	5
$?^*$	7	4	0	0	0	0	4	-5	-5	-5	-5	-5	-5	-6^*	-6	-6
@x	8	5	0	0	0	0	0	-5	-6	-6	-6	-6	-6	-6°	-7^*	-7
$?^*$	9	5	0	0	0	0	0	5	-6	-7	-7	-7	-7	-7	-7°	-8
M	10	6	0	0	0	0	0	0	6	7	7	-8	8	8	8	$-8^\circledast$

2.4.2.1 Adding Distance-Based Constraints

The evaluation of distance constraints $\mathcal{D}$ inside a pattern query $\mathcal{Q}$ is performed as a post filtering step after the pattern $\mathcal{S}$ evaluation. The *DPP* algorithm can only use a brute force approach since it maintains a trajectory as a sequence of regions but loses the spatial properties of these regions. Therefore, the *DPP* algorithm can only compute the distance for the constraint as a final step.

2.5 Experimental Evaluation

We run various experiments with real world and synthetic datasets to test the behavior of each technique under different settings. All experiments were run on an Intel Pentium-4 2.6 GHz processor running Linux 2.6.22 with 1 GB main memory. All implementations used the same disk manager framework with disk page size set to 4 KB for each index (*region-list* and *trajectory-list* indexes) and 16 KB for the raw trajectory archive.

For comparison purposes, we examined two previous pattern matching approaches. In particular, we modified the proposal in [31] and in [30], which we call here *Extended-KMP* (*E-KMP*) and *Extended-NFA* (*E-NFA*), respectively, and implemented them in our proposed FlexTrack system in order to fairly compare them against the *IJP* and *DPP* algorithms. The *E-KMP* contains extensions to handle the variable predicates ($?^*$, $?^+$) as well as the implicit/explicit temporal constraints. The *NFA* used in [30] finds simple event patterns in streaming data. Hence it is not formulated to evaluate topological relations or temporal constraints, as described in Sect. 2.2. We thus extend it to cover these as well, as to process queries with variables. To this end, a stack is created for each variable $@x$. If a variable appears in the query many times, a post processing check is performed at the accept state of the NFA. For fairness, all algorithms were tested using the same index framework, i.e., the *E-KMP* and the *E-NFA* algorithms receive a candidate set of trajectories similar to the *DPP* approach.

For real datasets, we use the *Truck* and *Buses* trajectorial data from [62]. Both datasets represent moving objects in the metropolitan area of Athens, Greece. The *Truck* dataset has 112,203 moving object locations generated from 276 trajectories where the longest trajectory timestamp is 13,540 timestamps. The *Buses* contains 66,096 moving object locations obtained from 145 trajectories of school buses with maximum timestamp 992. For simplicity of the experimental evaluation, we do not use real regions; instead we assume that the spatial domain (i.e., area of Athens) is partitioned into artificial regions using a uniform grid. These grid cells become the alphabet for our queries; hence in the rest the terms "region" and "cell" have the same meaning. To examine the effect of the alphabet size on the index structures we experiment with grid granularity starting from 25×25 up to 100×100.

We take this opportunity to note that in practice the region partition (i.e., the alphabet) depends only on the application and it is fixed. For example, if the user is interested in posing very detailed queries (e.g., street level) a finer partition should be used. We expect that in a real implementation one may use a fine grid partition at the lower level and (coarser) regions that the user understands at a higher level. Then an interesting problem is what size of grid partition will optimize the query response at the region level. Since the coverage of a region by grid cells may not be exact, false positives can be returned by the query, thus affecting query performance. This however is an orthogonal problem and is not addressed here; hence in the rest the terms "region" and "cell" have the same meaning.

For the synthetic datasets, we generated datasets of moving object trajectories. The dataset represents the freeway network of Indiana and Illinois states together. The 2-dimensional spatial universe is 1,000 miles long in each direction and contains up to 200,000 objects. Objects start at random positions on predefined routes in a road network and follow a Normal distribution with mean 60 mph and standard deviation 15 mph. We run simulations for 500 min (timestamps). For these datasets, the spatial universe was partitioned with a grid 100×100.

In order to generate relevant pattern queries, we randomly sample and fragment 100 trajectories. The length and location of each fragment are randomly chosen. These fragments are then concatenated to create a query. For the synthetic datasets we used pattern length from 5 up to 10 predicates. We also generated sets of pattern queries with different number of variable predicates (from 0 to 2). The location of each variable inside the query was randomly chosen. For queries with two variables, half of the patterns have the same variable twice, and the other half use 2 different variables (i.e., $@x$ and $@y$). For each experiment, we measured the average running time (in seconds) and the average I/O for a set of 100 queries. The query cost shown consists of the CPU time and the I/O time.

2.5.1 Evaluating Queries with Spatial Predicates

The first experiment, shown in Table 2.2, evaluates the total time (in seconds) required to execute four complex pattern queries on the *Buses* and *Truck* datasets. Since in the real datasets objects move in relatively similar ways, we experimented with larger number of predicates so as to create more selective queries. Moreover, queries S_1–S_4 contain between 2 and 4 variables and several wild-cards $?^+$ and $?^*$. The total number of predicates is specified by $|S|$, the number of fixed predicates is $|S_f|$, the number of trajectories returned is shown under $|A|$.

The results show that the *E-NFA* algorithm performs worse for all queries. This is because it cannot take advantage of the existing indexing structures so as to focus the search only on those parts of the trajectories that might contain answer (except from the original trajectory pruning using the region-list intersection). This is to be expected since the method has been designed for identifying patterns over streaming (non-archived) data. We experienced a similar behavior with the other real and also the synthetic datasets; hence we remove the *E-NFA* method from the following com-

Table 2.2 Query time (s) for real datasets

P	Dataset	$|S|$	$|S_f|$	$|\mathcal{A}|$	E-NFA	E-KMP	DPP	IJP
S_1	Buses	10	3	57	2.46	1.90	1.11	1.53
S_2	Buses	20	7	29	89.62	62.75	28.99	3.03
S_3	Trucks	20	7	76	111.91	54.68	30.28	10.57
S_4	Trucks	46	29	11	3.06	0.73	0.22	1.56

Fig. 2.8 Total number of index records for different alphabet sizes

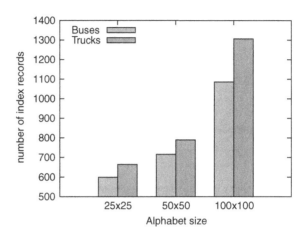

parisons. Among the rest, the *DPP* and *IJP* algorithms have typically more robust behavior; nevertheless, *E-KMP* still shows competitive behavior for some queries.

To examine the effect of the size of the alphabet on the index size, we experimented with the real datasets and different alphabets (by changing the grid size). As expected the increased number of letters in the alphabet increases the size of the index (see Fig. 2.8). Each trajectory visits more regions (which have smaller size) during its lifetime and thus generates more records in the index structure. Note that in this experiment, the size of index was very small compared to the raw data size (varying between 4 and 6 % for the *Buses* and 2 and 5 % for the *Trucks* dataset). The number of I/Os during the query evaluation however stays the same because each predicate in the query still corresponds to a single (though smaller) grid cell. As a result, the observed query times remain similar to the ones shown in Table 2.2.

To further examine the performance of the *DPP*, *IJP* and *E-KMP* algorithms we use the synthetic datasets. The next experiment evaluates the average running time required to execute 100 pattern queries while varying the number of spatial predicates in $\mathcal{P}$ from 5 to 10. It uses a synthetic dataset with 50,000 trajectories. The results appear in Fig. 2.9 (in log-scale) and for patterns with (a) no variable, (b) 1 variable, and (c) 2 variables.

As observed from these experiments, when increasing the number of predicates in the pattern, the query time of the *DPP* and *E-KMP* algorithms increases. For the *DPP* the larger pattern implies a larger matrix and thus more processing.

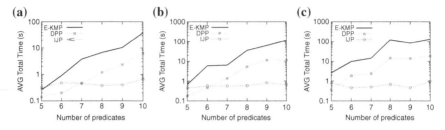

Fig. 2.9 Query time (s) when increasing the number of patterns in $\mathcal{P}$. (**a**) No variable, (**b**) 1 variable, (**c**) 2 variable

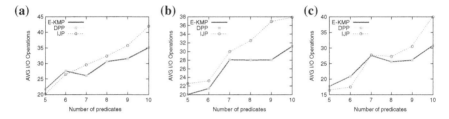

Fig. 2.10 Query I/O when increasing the number of patterns in $\mathcal{P}$. (**a**) No variable, (**b**) 1 variable, (**c**) 2 variable

The *E-KMP* is very sensitive to the number of ?* in the query; as the pattern increases in size the probability of more ?* increases (this effect will be examined further later). Nevertheless, the *DPP* algorithm is always more efficient than the *E-KMP* (typically by an order of magnitude).

The *IJP* algorithm is affected the least by the number of predicates. This is because processing in the *IJP* algorithm is guided by the *region-lists* of the first few predicates in the pattern (for example, the third list is accessed after a match in the first two lists is found, etc.). Hence, adding more lists does not directly affect the processing. As more predicates are added, the processing of the *E-KMP* and *DPP* starts increasing making the *IJP* a faster solution.

For the same experiments, Fig. 2.10 depicts the average I/O's for (a) 0 variable, (b) 1 variable, and (c) 2 variables. In particular *E-KMP* and *DPP* have identical I/O behavior since they are using the same approach to pick candidate *trajectory-lists* (without using the time constraints). Even though all three algorithms use the same indexes to retrieve objects, the *IJP* uses a different strategy (as described in Sect. 2.4) which results in a different I/O behavior. Nevertheless, all algorithms have comparable I/O behavior, leading us to the conclusion that the major differences in the overall processing time among the algorithms are not I/O based but mainly CPU bound.

We also performed experiments comparing the proposed index structure with R-trees. The R-tree was outperformed by our simpler grid structure. Since R-trees are data-driven structures the overlapping implies that several sub-trees need to be

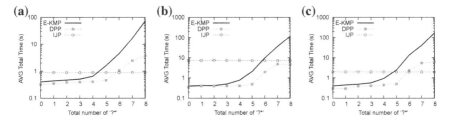

Fig. 2.11 Average running time versus number of ?* in $\mathcal{P}$. (**a**) No variable, (**b**) 1 variable, (**c**) 2 variable

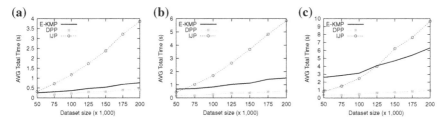

Fig. 2.12 Average running time versus number of trajectories. (**a**) No variable, (**b**) 1 variable, (**c**) 2 variable

analyzed. Furthermore, when MBRs over-approximate regions, the verification step at the end of the algorithm had to process significant amount of false positives.

2.5.2 Scalability Experiments

2.5.2.1 Increasing the Number of Wild-Cards in a Pattern

We next examined the performance of the three algorithms when varying the number of ?* wild-cards in the pattern. For these experiments, we randomly sampled 100 trajectories from the previous synthetic dataset and then extracted pattern queries of length 10. These patterns contain only the "." sequencer (i.e., no ?*). Using this query set we created a new set that has queries with one ?*. This set was created by randomly replacing one "." by a ?*. We continued in the same way creating a new query set with queries having two ?* by replacing an additional ".", etc. Figure 2.11 shows the average running time (in log-scale) to execute 100 queries varying the total number of ?* (from 0 to 8) in each pattern with (1) no variable, (b) 1 variable, and (c) 2 variables.

Again we observed that in all experiments the *DPP* algorithm is always faster than the *E-KMP*. As the number of ?* increases, the performance of *E-KMP* deteriorates drastically, showing the dependence of *E-KMP* to the ?* wild-cards. This is because each such wild-card forces the *E-KMP* approach to run more, shorter queries. More queries add to the processing time but also since these queries are smaller, the shifting function of *E-KMP* is not as effective. The *DPP* is up to four times faster than the *E-KMP* when there are eight wild-cards. For the *DPP*, the total processing time increases because more matches qualify as an answer. The performance of the *IJP* algorithm is independent of the number of the ?* wild-cards, since they are evaluated in the same way as the "." sequencers. As a result, as more wild-cards appear in the query, *IJP* will eventually become faster than the *DPP*.

2.5.2.2 Increasing the Number of Trajectories

We then varied the indexed dataset size to examine the scalability of the proposed algorithms. For these experiments, we used a synthetic dataset of 200,000 trajectories. We started with inserting the first 50,000 trajectories in the indexes and measured the query time (for an average of 100 queries each with five predicates, including 0, 1 and 2 variables). We repeated the experiment after adding an additional 25,000 trajectories. This incremental process continued with increments of 25,000 trajectories until the total of 200,000 trajectories in the archive.

The behavior of all algorithms grows linearly with the dataset size, as shown in Fig. 2.12. Recall that from our complexity analysis, both the *IJP* and *DPP* algorithms are proportional to the number of candidate trajectories; as more trajectories are added, this number increases thus affecting the overall performance accordingly. Again, the *DPP* algorithm behaves consistently better than the *E-KMP*. Among all algorithms, the *IJP* has the faster rate of increase. This is because, the larger datasets create large *region-lists* which directly affects the join processing cost. Moreover, *IJP* performs two join operations (one in the *region-lists* and one in the *variable-lists*) and both of them are directly affected by the size of the lists.

2.5.3 Evaluating Patterns with Spatial Predicates and Nearest Neighbors

We also performed experiments to examine how the algorithms behave when adding nearest neighbor predicates (i.e., pattern queries that contain both $\mathcal{P}$ and $\mathcal{D}$). We examined four query datasets varying the number of distance terms from one to four. Each distance term uses two variables (i.e., it is of the form $d(@x, @y)$ which corresponds to the very processing demanding *NN* query). All variables in each pattern query are different and their positions were randomly chosen. Figure 2.13 shows the results for queries using ten predicates while increasing the number of distance terms. Clearly, the *IJP* approach outperforms the "brute-force" approach of the *DPP* (up to two orders of magnitude). This is because *IJP* maintains the spatial properties of trajectories and can thus reuse the variable lists to avoid examining all possible bindings.

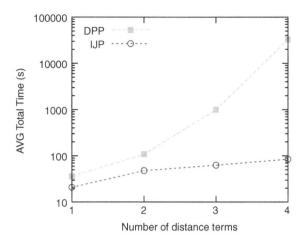

Fig. 2.13 Average running time versus number of distance terms in $\mathcal{D}$

2.5.4 Experimental Summary

In all our experiments the previous *E-KMP*-based approach (even optimized to use indexes) was outperformed by the *DPP* algorithm. Furthermore, its performance deteriorates drastically as the number of ?* wild-cards increases. Similarly, the *E-NFA* was outperformed by all algorithms. When comparing our two new algorithms, we observed the following:

1. For a small number of predicates the *DPP* algorithm is faster than the *IJP* algorithm. This is because the matrix M is small and thus it is processed very fast;
2. For larger number of predicates the *IJP* algorithm becomes faster since its performance is not affected by the increase in predicates, while the *DPP* is affected by the increase in the matrix size;
3. On the other hand, *IJP* is a join-based algorithm, hence the larger the dataset, the more expensive is the join step;
4. Nevertheless, *IJP* has more robust performance when considering distance-based queries (*NN*) as well, while the *DPP* (and *E-KMP*) algorithm needs to use a very consuming "brute-force" approach.

2.6 Final Remarks

In this chapter, we introduced the FlexTrack system for processing *flexible pattern queries* over trajectory archives. Such queries combine the ability of fixed and variable predicates, with explicit or implicit temporal constraints and distance-based constraints. Previous works have considered only subsets of FlexTrack and are based on variations of the *KMP* algorithm or use finite automata (*NFA*). We introduced two query processing techniques: one based on merge joins (*IJP*) and one based on subsequence matching (*DPP*). The experimental evaluation shows that our techniques improve substantially even over *optimized* (using indexing and preprocessing techniques) *KMP* and *NFA* approaches. Among our approaches, *IJP* is more robust in that it can easily support *NN* queries, while *DPP* is better for patterns with smaller number of predicates or wild-cards. Since however both approaches use the same indexing schemes, they can both be available to the user. As a future research topic, one could explore cost models that will enable a query optimizer to pick the best technique based on the query parameters (size of the pattern query, number of variables, wild-cards, etc). Another interesting topic of research is to extend the FlexTrack system to support complex pattern queries in GPUs (Graphics Processing Units), following a similar approach of using FPGAs (Field Programmable Gate Arrays) to evaluate pattern queries [63].

References

1. NAVCEN, U.C.G.N.C.: Navstar GPS User Equipment Introduction. http://www.navcen.uscg. gov/pubs/gps/gpsuser/gpsuser.pdf (1996)
2. Consumer & Governmental Affairs Bureau, F.: Wireless 911 services. http://www.fcc.gov/ guides/wireless-911-services
3. AccuTracking Inc.: AccuTracking. http://www.accutracking.com (2012)
4. iSECUREtrac: tracNET24. http://www.isecuretrac.com (2012)
5. Path Intelligence Inc.: FootPath. http://www.pathintelligence.com (2011)
6. Instedd Inc.: GeoChat. http://www.instedd.org/technologies/geochat/
7. Ijeh, A., Brimicombe, A., Preston, D., Imafidon, C.: Geofencing in a security strategy model. In: H. Jahankhani, A. Hessami, F. Hsu (eds.) Global Security, Safety, and Sustainability, *Communications in Computer and Information Science*, vol. 45, pp. 104–111. Springer, Berlin Heidelberg (2009). DOI http://dx.doi.org/10.1007/978-3-642-04062-7_11
8. Kollios, G., Papadopoulos, D., Gunopulos, D., Tsotras, V.J.: Indexing mobile objects using dual transformations. The VLDB Journal **14**(2), 238–256 (2005). DOI http://dx.doi.org/10. 1007/s00778-004-0139-z
9. Pelanis, M., Saltenis, S., Jensen, C.S.: Indexing the past, present, and anticipated future positions of moving objects. ACM Trans. Database Syst. **31**(1), 255–298 (2006). DOI http://dx. doi.org/10.1145/1132863.1132870
10. Tao, Y., Faloutsos, C., Papadias, D., Liu, B.: Prediction and indexing of moving objects with unknown motion patterns. In: Proceedings of the ACM SIGMOD International Conference on Management of Data, pp. 611–622. ACM (2004). DOI http://dx.doi.org/10.1145/1007568. 1007637
11. Tao, Y., Papadias, D.: Time-parameterized queries in spatio-temporal databases. In: Proceedings of the ACM SIGMOD International Conference on Management of Data, pp. 334–345. ACM (2002). DOI http://dx.doi.org/10.1145/564691.564730
12. Tao, Y., Sun, J., Papadias, D.: Analysis of predictive spatio-temporal queries. ACM Trans. Database Syst. pp. 295–336 (2003). DOI http://dx.doi.org/10.1145/958942.958943
13. Hadjieleftheriou, M., Kollios, G., Tsotras, V.J., Gunopulos, D.: Indexing spatiotemporal archives. The VLDB Journal **15**(2), 143–164 (2006). DOI http://dx.doi.org/10.1007/s00778-004-0151-3

14. Pfoser, D., Jensen, C.S., Theodoridis, Y.: Novel approaches in query processing for moving object trajectories. In: Proceedings of the International Conference on Very Large Data Bases (VLDB), pp. 395–406 (2000)
15. Benetis, R., Jensen, S., Karciauskas, G., Saltenis, S.: Nearest and reverse nearest neighbor queries for moving objects. The VLDB Journal **15**(3), 229–249 (2006). DOI http://dx.doi.org/10.1007/s00778-005-0166-4
16. Aggarwal, C.C., Agrawal, D.: On nearest neighbor indexing of nonlinear trajectories. In: Proceedings of the ACM SIGMOD-SIGACT-SIGART Symposium on Principles of Database Systems (PODS), pp. 252–259. ACM (2003). DOI http://dx.doi.org/10.1145/773153.773178
17. Ferhatosmanoglu, H., Stanoi, I., Agrawal, D., Abbadi, A.E.: Constrained nearest neighbor queries. In: Proceedings of the International Symposium on Advances in Spatial and Temporal Databases (SSTD), *Lecture Notes in Computer Science*, vol. 2121, pp. 257–278. Springer (2001). DOI http://dx.doi.org/10.1007/3-540-47724-1_14
18. Cudre-Mauroux, P., Wu, E., Madden, S.: Trajstore: An adaptive storage system for very large trajectory data sets. In: Proceedings of the IEEE International Conference on Data Engineering (ICDE), pp. 109–120. IEEE Computer Society (2010). DOI http://dx.doi.org/10.1109/ICDE.2010.5447829
19. Lee, J.G., Han, J., Whang, K.Y.: Trajectory clustering: a partition-and-group framework. In: Proceedings of the ACM SIGMOD International Conference on Management of Data, pp. 593–604. ACM (2007). DOI http://dx.doi.org/10.1145/1247480.1247546
20. Mamoulis, N., Cao, H., Kollios, G., Hadjieleftheriou, M., Tao, Y., Cheung, D.W.: Mining, indexing, and querying historical spatiotemporal data. In: Proceedings of the ACM SIGKDD International Conference on Knowledge Discovery and Data Mining, pp. 236–245. ACM (2004). DOI http://dx.doi.org/10.1145/1014052.1014080
21. Papadias, D., Tao, Y., Zhang, J., Mamoulis, N., Shen, Q., Sun, J.: Indexing and retrieval of historical aggregate information about moving objects. IEEE Data Eng. Bull. (2002)
22. Papadias, D., Shen, Q., Tao, Y., Mouratidis, K.: Group nearest neighbor queries. In: Proceedings of the IEEE International Conference on Data Engineering (ICDE), pp. 301–312. IEEE Computer Society (2004). DOI http://dx.doi.org/10.1109/ICDE.2004.1320006
23. Bakalov, P., Hadjieleftheriou, M., Tsotras, V.J.: Time relaxed spatiotemporal trajectory joins. In: Proceedings of the ACM SIGSPATIAL International Conference on Advances in Geographic Information Systems, pp. 182–191. ACM (2005). DOI http://dx.doi.org/10.1145/1097064.1097091
24. Arumugam, S., Jermaine, C.: Closest-point-of-approach join for moving object histories. In: Proceedings of the IEEE International Conference on Data Engineering (ICDE), pp. 86–86. IEEE Computer Society (2006). DOI http://dx.doi.org/10.1109/ICDE.2006.36
25. Vieira, M.R., Bakalov, P., Tsotras, V.J.: Querying trajectories using flexible patterns. In: Proceedings of the International Conference on Extending Database Technology (EDBT), pp. 406–417. ACM (2010). DOI http://dx.doi.org/10.1145/1739041.1739091
26. Vieira, M.R., Bakalov, P., Tsotras, V.J.: FlexTrack: A system for querying flexible patterns in trajectory databases. In: Proceedings of the International Symposium on Advances in Spatial and Temporal Databases (SSTD), *Lecture Notes in Computer Science*, vol. 6849, pp. 475–480. Springer, Berlin Heidelberg (2011). DOI http://dx.doi.org/10.1007/978-3-642-22922-0_34
27. Esri: ArcGIS. http://www.esri.com (2013)
28. Sadri, R., Zaniolo, C., Zarkesh, A., Adibi, J.: Expressing and optimizing sequence queries in database systems. ACM Trans. Database Syst. **29**(2), 282–318 (2004). DOI http://dx.doi.org/10.1145/1005566.1005568
29. Seshadri, P., Livny, M., Ramakrishnan, R.: SEQ: A model for sequence databases. In: Proceedings of the IEEE International Conference on Data Engineering (ICDE), pp. 232–239. IEEE Computer Society (1995). DOI http://dx.doi.org/10.1109/ICDE.1995.380388
30. Agrawal, J., Diao, Y., Gyllstrom, D., Immerman, N.: Efficient pattern matching over event streams. In: Proceedings of the ACM SIGMOD International Conference on Management of Data, pp. 147–160. ACM (2008). DOI http://dx.doi.org/10.1145/1376616.1376634

31. du Mouza, C., Rigaux, P., Scholl, M.: Efficient evaluation of parameterized pattern queries. In: Proceedings of the ACM International Conference on Information and Knowledge Management (CIKM), pp. 728–735. ACM (2005). DOI http://dx.doi.org/10.1145/1099554.1099731
32. Erwig, M., Schneider, M.: Spatio-temporal predicates. IEEE Trans. on Knowl. and Data Eng. **14**(4), 881–901 (2002). DOI http://dx.doi.org/10.1109/TKDE.2002.1019220
33. Hadjieleftheriou, M., Kollios, G., Bakalov, P., Tsotras, V.J.: Complex spatio-temporal pattern queries. In: Proceedings of the International Conference on Very Large Data Bases (VLDB), pp. 877–888 (2005)
34. Sakr, M.A., Güting, R.H.: Spatiotemporal pattern queries in SECONDO. In: Proceedings of the International Symposium on Advances in Spatial and Temporal Databases (SSTD), *Lecture Notes in Computer Science*, vol. 5644, pp. 422–426. Springer, Berlin Heidelberg (2009). DOI http://dx.doi.org/10.1007/978-3-642-02982-0_32
35. Tao, Y., Papadias, D., Shen, Q.: Continuous nearest neighbor search. In: Proceedings of the International Conference on Very Large Data Bases (VLDB), pp. 287–298 (2002)
36. Agarwal, P.K., Arge, L., Erickson, J.: Indexing moving points. In: Proceedings of the ACM SIGMOD-SIGACT-SIGART Symposium on Principles of Database Systems (PODS), pp. 175–186. ACM (2000). DOI http://dx.doi.org/10.1145/335168.335220
37. Cai, M., Revesz, P.: Parametric R-tree: An index structure for moving objects. In: COMAD, pp. 57–64 (2000)
38. Elbassioni, K.M., Elmasry, A., Kamel, I.: An efficient indexing scheme for multi-dimensional moving objects. In: International Conference on Database Theory (ICDT), pp. 425–439. Springer-Verlag (2003). DOI http://dx.doi.org/10.1007/3-540-36285-1_28
39. Jensen, C.S., Lin, D., Ooi, B.: Query and update efficient B+-Tree based indexing of moving objects. In: Proceedings of the International Conference on Very Large Data Bases (VLDB), pp. 768–779 (2004)
40. Patel, J.M., Chen, Y., Chakka, V.P.: Stripes: an efficient index for predicted trajectories. In: Proceedings of the ACM SIGMOD International Conference on Management of Data, pp. 635–646. ACM (2004). DOI http://dx.doi.org/10.1145/1007568.1007639
41. Prabhakar, S., Xia, Y., Kalashnikov, D.V., Aref, W.G., Hambrusch, S.E.: Query indexing and velocity constraint indexing: Scalable techniques for continuous queries on moving objects. IEEE Trans. on Computers pp. 1–17 (2002). DOI http://dx.doi.org/10.1109/TC.2002.1039840
42. Saltenis, S., Jensen, C.S.: Indexing of moving objects for location-based services. In: Proceedings of the IEEE International Conference on Data Engineering (ICDE), pp. 463–472. IEEE Computer Society (2002). DOI http://dx.doi.org/10.1109/ICDE.2002.994759
43. Tao, Y., Papadias, D.: MV3R-Tree: A spatio-temporal access method for timestamp and interval queries. In: Proceedings of the International Conference on Very Large Data Bases (VLDB), pp. 431–440 (2001)
44. Tao, Y., Papadias, D., Sun, J.: The TPR*-tree: An optimized spatio-temporal access method for predictive queries. In: Proceedings of the International Conference on Very Large Data Bases (VLDB), pp. 790–801. VLDB Endowment (2003)
45. Mokbel, M.F., Aref, W.G.: SOLE: scalable on-line execution of continuous queries on spatio-temporal data streams. The VLDB Journal**17**(5), 971–995 (2008). DOI http://dx.doi.org/10.1007/s00778-007-0046-1
46. Anagnostopoulos, A., Vlachos, M., Hadjieleftheriou, M., Keogh, E.J., Yu, P.S.: Global distance-based segmentation of trajectories. In: Proceedings of the ACM SIGKDD International Conference on Knowledge Discovery and Data Mining, pp. 34–43. ACM (2006). DOI http://dx.doi.org/10.1145/1150402.1150411
47. Cai, Y., Ng, R.: Indexing spatio-temporal trajectories with Chebyshev polynomials. In: Proceedings of the ACM SIGMOD International Conference on Management of Data, pp. 599–610. ACM (2004). DOI http://dx.doi.org/10.1145/1007568.1007636
48. Ni, J., Ravishankar, C.V.: PA-Tree: A parametric indexing scheme for spatio-temporal trajectories. In: Proceedings of the International Symposium on Advances in Spatial and Temporal

Databases (SSTD), *Lecture Notes in Computer Science*, vol. 3633, pp. 254–272. Springer-Verlag (2005). DOI http://dx.doi.org/10.1007/11535331_15
49. Lee, S.L., Chun, S.J., Kim, D.H., Lee, J.H., Chung, C.W.: Similarity search for multidimensional data sequences. In: Proceedings of the IEEE International Conference on Data Engineering (ICDE), pp. 599–608. IEEE Computer Society (2000). DOI http://dx.doi.org/10.1109/ICDE.2000.839473
50. Yanagisawa, Y., Akahani, J.i., Satoh, T.: Shape-based similarity query for trajectory of mobile objects. In: M.S. Chen, P. Chrysanthis, M. Sloman, A. Zaslavsky (eds.) Mobile Data Management, *Lecture Notes in Computer Science*, vol. 2574, pp. 63–77. Springer, Berlin Heidelberg (2003). DOI http://dx.doi.org/10.1007/3-540-36389-0_5
51. Vlachos, M., Kollios, G., Gunopulos, D.: Discovering similar multidimensional trajectories. In: Proceedings of the IEEE International Conference on Data Engineering (ICDE), pp. 673–684. IEEE Computer Society (2002). DOI http://dx.doi.org/10.1109/ICDE.2002.994784
52. IBM: Informix. http://www.ibm.com/software/data/informix/ (2012)
53. Mokhtar, H., Su, J., Ibarra, O.: On moving object queries. In: Proceedings of the ACM SIGMOD-SIGACT-SIGART Symposium on Principles of Database Systems (PODS), pp. 188–198. ACM (2002). DOI http://dx.doi.org/10.1145/543613.543638
54. Sakr, M.A., Güting, R.H.: Spatiotemporal pattern queries. GeoInformatica **15**(3), 497–540 (2011). DOI http://dx.doi.org/10.1007/s10707-010-0114-3
55. Qu, Y., Wang, C., Gao, L., Wang, X.S.: Supporting movement pattern queries in user-specified scales. IEEE Trans. on Knowl. and Data Eng. **15**(1), 26–42 (2003). DOI http://dx.doi.org/10.1109/TKDE.2003.1161580
56. Laube, P., Imfeld, S., Weibel, R.: Discovering relative motion patterns in groups of moving point objects. J. of Geog. Inf. Science **19**(6), 639–668 (2005). DOI http://dx.doi.org/10.1080/13658810500105572
57. Knuth, D.E., Jr., J.H.M., Pratt, V.R.: Fast pattern matching in strings. SIAM J. Comput. **6**(2), 323–350 (1977). DOI http://dx.doi.org/10.1145/1146809.1146812
58. Nievergelt, J., Hinterberger, H., Sevcik, K.C.: The grid file: An adaptable, symmetric multikey file structure. ACM Trans. Database Syst. **9**(1), 38–71 (1984). DOI http://dx.doi.org/10.1145/348.318586
59. Robinson, J.T.: The K-D-B-tree: a search structure for large multidimensional dynamic indexes. In: Proceedings of the ACM SIGMOD International Conference on Management of Data, pp. 10–18. ACM (1981). DOI http://dx.doi.org/10.1145/582318.582321
60. Freeston, M.: The BANG file: A new kind of grid file. In: Proceedings of the ACM SIGMOD International Conference on Management of Data, pp. 260–269. ACM (1987). DOI http://dx.doi.org/10.1145/38713.38743
61. Fagin, R., Lotem, A., Naor, M.: Optimal aggregation algorithms for middleware. In: Proceedings of the ACM SIGMOD-SIGACT-SIGART Symposium on Principles of Database Systems (PODS), pp. 102–113. ACM (2001). DOI http://dx.doi.org/10.1145/375551.375567
62. The Chorochronos Archive: The R Tree Portal. http://www.chorochronos.org (2013)
63. Moussalli, R., Vieira, M.R., Najjar, W.A., Tsotras, V.J.: Stream-mode FPGA acceleration of complex pattern trajectory querying. In: Proceedings of the International Symposium on Advances in Spatial and Temporal Databases (SSTD), *Lecture Notes in Computer Science*, vol. 8098, pp. 201–222. Springer (2013). DOI http://dx.doi.org/10.1007/978-3-642-40235-7_12

Chapter 3
Pattern Queries for Mobile Phone-Call Databases

3.1 Introduction

The recent adoption of ubiquitous computing technologies by very large portions of the world population has enabled, for the first time in human history, to capture large scale spatio-temporal data about human motion. In this context, mobile phones play a key role as sensors of human behavior since they are typically owned by one individual that carries it at (almost) all times and are nearly ubiquitously used. Hence, it is no surprise that most of the quantitative data about human motion has been gathered via Call Detail Records (CDRs) of cell phone networks.

When a cell phone makes/receives a phone call the information regarding the call is logged in the form of a CDR. This information includes, among others, originating and destination phone numbers, the time and date when the call started, and the towers used, which gives an approximation of the caller's/callee's geographical location. Such data is very rich and has been used recently for several applications, such as to study user's social networks [1–3], human mobility behaviors [4, 5], and cellular network improvement [6].

The volume of data generated by a given operator in the form of CDRs is huge, and it contains valuable spatio-temporal information at different levels of granularity (e.g. citywide, statewide, and nationwide). This information is relevant not only for telecommunication operators but also as a base for a broader set of applications with social connotations like commuting patterns, transportation routes, concentrations of people, modeling of virus spreading, etc. The ability to efficiently query CDR databases to search for spatio-temporal patterns is key to the development of such applications. Nevertheless, the commercial systems available cannot efficiently handle this kind of spatio-temporal processing. One possible solution to search for such patterns is to perform a sequential scanning of the *entire* CDR database and, for each user, check whether it qualifies using a subsequence matching-like algorithm (e.g. Knuth-Morris-Pratt (KMP) [7]). Such naive approach however is computationally extremely expensive due to the amount of users/CDRs to be processed. Furthermore, there is the fact that no information about the temporal dimension of the pattern (e.g.

M. R. Vieira and V. J. Tsotras, *Spatio-Temporal Databases*,
SpringerBriefs in Computer Science, DOI: 10.1007/978-3-319-02408-0_3,
© The Author(s) 2013

within given time frame) or spatial properties (e.g. in a given neighborhood) can be specified.

Taking into consideration the large volume of data and current implementation of the CDR storage systems for telecommunication providers, one effective way to support such spatio-temporal pattern queries is to extend the current systems with some indexes and algorithms to efficiently process such queries. One aspect that has to be considered is that commercial storage systems are in their majority implemented on top of Relational Database Management System (RDBMS). Therefore the provided solution should use the available RDBMS infrastructure such as tables, indexes (e.g. inverted indexes and B-trees), merge-join algorithms, and so on.

In this chapter, we present the Spatio-Temporal Pattern System (STPS) to query spatio-temporal patterns in CDR databases. The STPS allows users to express mobility pattern queries with a regular expression-like language that can include *variables* in the pattern specification. Variables serve as "placeholders" in the pattern for *explicit* spatial regions and their value is determined during the pattern query evaluation. An example for a query with variables is the pattern "find users who visited the same mall twice in the last 24 hours". In this scenario we do not know in advance which one is the mall visited by the user. So we use variables which can take values from the set of malls to specify the user behavior in a pattern query. We have to pay attention that in the above example the variable should appear *twice* in the pattern.

STPS also includes lightweight index structures that can be easily implemented in most commercially RDBMS. We present an extensive experimental evaluation of the proposed techniques using two large, real-world CDR databases. The experimental results reveal that the proposed STPS framework is scalable and efficient under several scenarios tested. Our proposed system is up to 1,000 times faster than a base line implementation, making the STPS a very robust approach for querying and analyzing very large phone-call databases.

This chapter presents a continuation of our previous work, described in Chap. 2, in pattern query evaluation in trajectorial archives. In this chapter we adopt that approach and study its application in the domain of CDR databases. In particular, we modified the join-based evaluation algorithm to handle trajectories specified in CDR format rather than the traditional form, defined as sequence of object locations with their *longitude* and *latitude* coordinates. This change in the data format poses changes in the query languages as well. In Chap. 2, the query language includes several query predicates that are well suited when the exact location of the object is known for a continuous period of time. An example of such a predicate is the *distance-based predicate* used to find trajectories that passed *as close as possible* to some area of interest. In a CDR database however, the exact location of the mobile user is unknown and users are not continuously monitored. Thus, the pattern language proposed here is more suitable for CDR databases (e.g. cells, user defined areas, temporal predicates to track *hopping* during a call or for different calls). The language proposed in this chapter also supports user defined constraints (e.g. conditions, inequalities, time constraints). Furthermore, the query evaluation system is redesigned to work with the features (e.g. tables, B^+-trees) of a commercially available RDBMS, since CDR databases are typically implemented in such systems.

The remainder of this chapter is organized as follows: Section 3.2 discusses the related work; Sect. 3.3 provides some basic descriptions on the infrastructure; Sect. 3.4 provides the formal description of the STPS language; the proposed system is described in Sect. 3.5 and its experimental evaluation appears in Sects. 3.6, 3.7 concludes the chapter with the final remarks.

3.2 Related Work

Infrastructures for querying spatio-temporal patterns have already been studied in the literature in different contexts, mainly for: (1) time-series databases; (2) similarity between trajectories; and (3) single predicate for trajectory data.

Pattern queries have been used in the past for querying time series using SQL-like query language [8, 9], or event streams using a NFA-based method [10]. Our work differs from those solutions mainly because it provides a richer language to specify spatio-temporal patterns and an efficient way to evaluate them. For moving object data, patterns have been examined in the context of query language and modeling issues [11, 12] as well as query evaluation algorithms [13].

Similarity search among trajectories has been also well studied. Work in this area focuses on the use of different distance metrics to measure the similarity between trajectories (e.g. [14–17]).

Single predicate queries for trajectory data, like Range and *NN* queries, have been well studied in the past (e.g. [18]). In these contexts, a query is expressed by a single range or *NN* predicate. To make the evaluation process more efficient, the query predicates are typically evaluated utilizing hierarchical spatio-temporal indexing structures [19]. Most structures use the concept of Minimum Bounding Regions (MBR) to approximate the trajectories, which are then indexed using traditional spatial access methods, like the MVR-tree [20]. These solutions, however, are focused only on single predicate queries and further constructions to build a more complex query, e.g. a sequence of combination of both predicates, are not supported. In [13] an incremental ranking algorithm for simple spatio-temporal pattern queries is presented. These queries consist of range and *NN* predicates specified using only *fixed* regions. Our work differs in that we provide a more general and powerful query framework where queries can involve both fixed and *variable* regions as well as topological operators, temporal predicates, constraints, etc., and an explicit ordering of the predicates along the temporal axis.

In [21] a *KMP*-based algorithm [7] is used to process patterns in trajectorial achieves. This work, however, focuses only on the *contain* topological predicate and cannot handle *explicit* or *implicit* temporal ordering of predicates. Furthermore, this approach on evaluating patterns is effectively reduced to a sequential scanning over the list of trajectories stored in the repository: each trajectory is checked individually, which becomes prohibitive for large trajectory archives. We show in Sect. 3.6 that this approach is very inefficient.

3.3 Background

Cell phone networks are built using a set of Base Transceiver Stations (BTS) that are in charge of communicating mobile phone devices with the cell network. The area covered by a BTS is called a cell. A BTS has one or more directional antennas (typically two or three, covering 180 or 120°, respectively) that define a sector and all the sectors of the same BTS define the cell. At any given moment in time, a cell phone is covered by one or more antennas. Depending on the network traffic, the phone selects the BTS to connect to. The geographical area covered by a cell depends mainly on the power of individual antennas. Depending on the population density, the area covered by a cell ranges from less than $1\,km^2$, in dense urban areas, to more than $5\,km^2$, in rural areas. Each BTS has latitude/longitude attributes that indicate its location, a unique identifier BTS_{id}, and the polygon representing its cell. For simplicity, we assume that the cell of each BTS is a 2-dimensional non-overlapping region, and we use Voronoi diagrams to define the covering areas of the set of BTSs considered. Figure 3.1a presents a set of BTSs with the original coverage for each cell, and Fig. 3.1b the simulated coverage obtained using Voronoi diagrams. While simple, this approach gives us a good approximation of the coverage area of each BTS. In practice, to build the real diagram of coverage, one has to consider several factors in the mobile network, e.g. power and orientation of each antenna.

CDR databases are populated when a mobile phone, connected to the network, makes/receives a phone call or uses a service in the network (e.g., SMS, MMS). In the process, the information regarding the time and the BTS where the user was located when the call was initiated is logged, which gives an *indication* of the user's geographical location at a given period in time. Note that no information about the *exact* user's location inside a cell is known. Furthermore, for a given call it is possible to store not only the initial BTS during the period of a call, but also all BTSs used during it in case caller/callee move to other cells in the network (*hopping*). The STPS supports this richer representation of the users' mobility.

The following attributes from CDR databases are used in the STPS system:

1. the originating phone number $phone_{id}^o$;

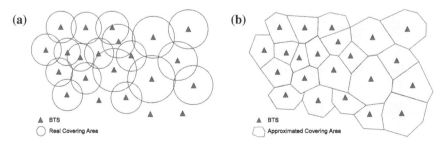

Fig. 3.1 Original and approaximated coverage areas, (**a**) Original coverage areas of a set of BTS, (**b**) Approximation of coverage areas by a Voronoi diagram

Table 3.1 A set of CDRs representing 4 different calls

Timestamp	Duration	Phone$_{id}^o$	Phone$_{id}^d$	BTS$_{id}^o$	BTS$_{id}^d$	Type
123001	3	324542	333434	231	121	V
123004	2	324542	333434	232	435	V
123006	5	324542	333434	234	121	V
123235	2	324542	334212	235	231	V
123237	4	324542	334212	231	233	V
124113	3	333434	324541	238	231	V
124116	4	333434	324541	239	231	V
124116	1	334212	333434	451	239	S

2. the destination phone number $phone_{id}^d$;
3. the type of service (voice: V, SMS: S, MMS: M, etc.);
4. the BTS identifier used by the originating number (BTS$_{id}^o$);
5. the BTS identifier used by the destination number (BTS$_{id}^d$);
6. *timestamp* (date/time) of the connection between $phone_{id}^o$ and $phone_{id}^d$ in BTS$_{id}^o$ and BTS$_{id}^d$, respectively; and
7. the duration *dur* while $phone_{id}^o$ and $phone_{id}^d$ connected to BTS$_{id}^o$ and BTS$_{id}^d$ (*hopping* enabled), respectively.

Since in the STPS we are only interested in users' mobility, we do not make any distinctions between caller and callee. Therefore, the superscript symbols (^{o} and ^{d}) in $phone_{id}$ and BTS$_{id}$ are omitted in the STPS language and framework. The BTS identifier is only known for $phone_{id}$ that are clients of the telecommunication operator keeping the CDR database. When the *hopping* is enabled, a new CDR row is created every time either users connects to different BTS$_{id}$ during the same phone call, otherwise, a single CDR is stored to represent the initial position of $phone_{id}^o$ and $phone_{id}^d$ for the total duration of the call *dur*.

Table 3.1 shows a set of CDRs for 4 distinct calls. In this example the BTS *hopping* option is enabled. Phone number 324542 makes a phone call to 333434 starting in BTS$_{id}^o = 231$ at timestamp 123212. Then the user 324542 moves from BTS$_{id}^o = 231$ to BTS$_{id}^o = 232$ 3 min after starting the call, generating another record in the database. After 2 min, user 324542 moves to BTS$_{id}^o = 234$ staying there for 5 min. The user 333434 is connected to BTS$_{id}^d = 121$, then to 435, and then back to 121 during the call. When a user is connected to a particular BTS$_{id}$, it does not necessary mean that the user is on the same place for the whole period of connection. The second call represents the call made from 324542 to 334212, and the third one from 333434 to 324541. The 8th entry of the table details an SMS sent from 334212 to 333434 when they were connected to BTS$_{id}^o = 451$ and BTS$_{id}^d = 239$, respectively. If the BTS *hopping* was not enabled, the first three entries would have been presented as a single one, with just the initial BTS$_{id}^o = 231$ and a total duration of 10 min.

3.4 The STPS Pattern Query Language

We define a trajectory $T(phone_{id})$ of a mobile user with identifier $phone_{id}$ in CDR databases as a sequence of records $\{\langle phone_{id}, \text{BTS}_{id}, t_1, dur_1\rangle, \ldots, \langle phone_{id}, \text{BTS}_{id}, t_m, dur_m\rangle\}$, where BTS_{id} is the BTS identifier which serviced the mobile user $phone_{id}$ at timestamp t_i for the duration of time dur_i ($t_i, t_m \in \mathbb{N}, t_i < t_m$ and $dur_i \in \mathbb{N}$). This trajectory definition covers both formats described in the previous section: (i) as a sequence of BTSs where the user was connected to the mobile network; or (ii) as a sequence of a trajectory *segments* (at a BTS level) where each *segment* represents the movement of the user between two BTS during a phone call. We assume that CDRs using this representation are stored in an archive as shown in Fig. 3.3d.

The STPS language uses the above definition of a trajectory to covers both data formats; i.e., we can query for patterns using records for the same phone call or different calls. This is achieved by associating temporal predicates for each spatial predicate which can be used to restrict the user movements into a time frame of a single phone call. In the following we describe in details the syntax of the STPS pattern query language and its components: the spatial predicates, the temporal predicates, and the set of spatio-temporal constraints.

3.4.1 The STPS Language Syntax

A pattern query $\mathcal{Q}$ is defined as $\mathcal{Q} = (\mathcal{S} \ [\bigcup \mathcal{C}])$, where $\mathcal{S}$ is a sequential pattern and $\mathcal{C}$ is an optional set of spatio-temporal constraints. The set of constraints $\mathcal{C}$ is used to specify spatial and/or temporal constraints that an answer has to satisfy in order to be considered as part of the result. A trajectory with identifier $phone_{id}$ matches the pattern query $\mathcal{Q}$ if it satisfies both the sequential pattern $\mathcal{S}$ and the set of spatio-temporal constraints $\mathcal{C}$. A sequential pattern $\mathcal{S}$ is defined as a sequence of an arbitrary number n of spatio-temporal predicates $\mathcal{S} = \{\mathcal{P}_1.\mathcal{P}_2., \cdots, .\mathcal{P}_n\}$.

Each spatio-temporal predicate $\mathcal{P}_i \in \mathcal{S}$ is defined by a triplet $\mathcal{P}_i = \langle op_i, \mathcal{R}_i[, t_i]\rangle$, where op_i represents a topological relationship operator, $\mathcal{R}_i$ a spatial region, and t_i the optional temporal predicate. The operator op_i describes the topological relationship that the spatial region $\mathcal{R}_i$ and the coverage area of the BTS defining a trajectory with identifier $phone_{id}$ must satisfy over the (optional) temporal predicate t_i. Figure 3.2 details formally the syntax of the STPS language.

3.4.2 Patterns with Spatial Predicates

A key part of our STPS language syntax is the definition of the spatial alphabet Σ, used in the spatio-temporal predicates $\mathcal{P}_i$. We choose the Voronoi diagram cells that

Fig. 3.2 The STPS Pattern
Query Language

$$\mathcal{Q} := (\mathcal{S} \; [\bigcup \mathcal{C}])$$
$$\mathcal{S} := \{\mathcal{P}_1.\mathcal{P}_2.,...,.\mathcal{P}_n\}, |\mathcal{S}| = n$$
$$\mathcal{P}_i := \langle op_i, \mathcal{R}_i[, t_i] \rangle$$
$$op_i := disjoint \mid meet \mid overlap \mid equal \mid inside \mid$$
$$contains \mid covers \mid coveredby$$
$$\mathcal{R}_i \in \{\Sigma \cup \Gamma\}$$
$$t_i := (t_{from} : t_{to}) \mid t_s \mid t_r$$

represent the covering areas of each BTS to serve as "letters" in our alphabet Σ. This is because the BTS coverage areas represent the finest level of granularity in which the data is stored in CDR databases. In the rest of the chapter we use capital letters to represent the set of BTS coverage areas in the system, e.g. $\Sigma = \{A, B, C, ...\}$. Such coverage areas can participate as spatial regions $\mathcal{R}_i$ in the definition of the spatio-temporal predicates $\mathcal{P}_i$.

The users however are not restricted to use only BTS coverage areas in their queries. On top of this BTS coverage partitioning the user can define its own geographical maps with different resolution and different types of regions (e.g., school districts, airports, and shopping). Also, users can define polygons defined by a set of latitude/longitude pairs to define a set of areas. All other regions, defined by the user, have to be approximated by set of coverage areas in the alphabet Σ. For instance, one can define the Airport area of a city by creating regions $Airport = \{D, E, H\}$ and $Stadium\text{-}1 = \{S1\}$, where the $Airport$ area is approximated by the union of the coverage areas of BTS D, E and H and the $Stadium\text{-}1$ is approximated by the coverage area of BTS $S1$. The same BTS_{id} can be used in the definition of multiple regions and not all BTS have to be included in each geographical map.

Inside the spatial predicates $\mathcal{P}_i$ we use finite set of spatial regions $\mathcal{R}_i$. Those regions can be one of the following: (i) a particular $BTS_{id} \in \Sigma$; (ii) an alias $\mathcal{A}$ defined by a set of one or more $BTS_{id} \in \Sigma$; or (iii) a variable in Γ. We refer to the first two groups of spatial regions $\mathcal{R}_i$ as *predefined* spatial regions. A predefined region (i.e., $S1 \in \Sigma$) is explicitly specified by the user in the query predicate (e.g. $Stadium\text{-}1 = \{S1\}$ in our example). In contrary, the third group of spatial regions, termed *variable* spatial regions, references an arbitrary region in the map and it is denoted by a lowercase letter preceded by the @ symbol (e.g. @x). A variable region is defined using symbols from the set $\Gamma = \{@a, @b, @c, ...\}$. Unless otherwise specified, a *variable* takes a single value (instance) from Σ (e.g. @$a=C$); however, in general, one can also specify in $\mathcal{C}$ the possible values of a specific *variable* as a subset of Σ (e.g., any city district with museums). Conceptually, *variables* work as placeholders for explicit spatial regions and can become instantiated (bound to a specific region) during the query evaluation in a process similar to unification in logical programming.

Moreover, the same *variable* @x can appear in several different predicates of pattern $\mathcal{S}$, referencing to the same region everywhere it occurs. This is useful for specifying complex queries that involve revisiting the same region many times. For example, a query like $@x.S1.@x$ finds mobile users that started from some region (denoted by variable $@x$), then at some point passed by region $S1$ and then they visited the same region they started from.

We finish with the description of the last component of the spatial predicate: the topological relationship operator op_i. In this work we use the eight topological relationships: *disjoint, meet, overlap, equal, inside, contains, covers* and *coveredby* defined by [11]. Given a phone user record $\langle phone_{id}, BTS_j, t_i \rangle$ and a region $\mathcal{R}_i$, the operator op_i returns a boolean value whether the coverage area in the phone user record BTS_j and the region $\mathcal{R}_i$ satisfy the topological relationship op_i (e.g., an *Inside* operator will return value *true* if the user associated with $phone_{id}$ was serviced by BTS which has coverage area inside the spatial region $\mathcal{R}_i$. For simplicity in the rest of the chapter we assume that the spatial operator is *Inside* and it is thus omitted from the query examples.

3.4.3 Patterns with Temporal Predicates

As it was mentioned in the definition of the STPS language a spatio-temporal predicate $\mathcal{P}_i$ may include an explicit temporal predicates t_i. Those predicates can be in the form of: (a) time *interval* $(t_{from} : t_{to})$ where $t_{from} \leq t_{to}$ (for example "between 4 pm and 5 pm"); (b) time *snapshot* t_s (for example "at 3:35 pm"); or (c) time *relative* $t_r = t_i - t_{i-1}$ from the time instance t_{i-1} when the previous spatio-temporal predicate $\mathcal{P}_{i-1}$ satisfied (for example "1 hour after the user left his home"). Those temporal predicates imply that the spatial relationship op_i between BTS_j and region $\mathcal{R}_i$ should be satisfied in the specified time frame t_i (e.g. "passed by area $S1$ between 4 pm and 5 pm"). If the temporal predicate is not specified, we assume that the spatial relationship can be satisfied *any time* in the duration of a call. For simplicity we assume that if two predicates $\mathcal{P}_i$, P_j occur within pattern $\mathcal{S}$ (where $i < j$) and have temporal predicates t_i, t_j, respectively, then these intervals do not overlap and t_i occurs before t_j on the time dimension.

3.4.4 Patterns with Spatio-Temporal Constraints

In order to restrict values that can be matched to spatio-temporal predicates, the STPS language supports an optional set of spatio-temporal constraints $\mathcal{C}$. To qualify a phone user has to first satisfy $\mathcal{S}$ and then $\mathcal{C}$. $\mathcal{C}$ works like a pos-filter to eliminate phone users that do not satisfy $\mathcal{C}$. Some examples of spatio-temporal constraints can be: $@x! = @y$, $@z = \{A, B, C\}$, $Period(t_i) =$ "Weekend", $Day(t_i) =$ "Wednesday", among many others.

3.4.5 STPS Language Example

We now provide a complete example of pattern using the STPS language. One example is: "find all mobile users that, on Saturdays, first start in an arbitrary area different to *Neighborhood-2* in the morning, then immediately went by *Airport*, then by the *Stadium-1* between 6 pm and 8 pm, then went in the *Neighborhood-1* neighborhood between 8pm and 10pm, and finally returned to their first area". This query example finds for mobile users that followed a pattern of movements where the first and last locations are not specified but have to be the same ($@x$); three other spatial predicates are defined over areas of interests; several temporal predicates are also defined; and finally spatio-temporal constraints are specified to filter out the results. This pattern query can be expressed in the STPS language as follows: $\mathcal{Q} := (\langle @x,$ $t_{from}=8\,am : t_{to}=3\,pm \rangle. \langle Airport, t_r=1\,min \rangle. \langle Stadium-1, t_{from}=6\,pm : t_{to}=8\,pm \rangle.$ $\langle Neighborhood-1,\ t_{from}=8\ pm\ :\ t_{to}=10\ pm \rangle. \langle @x \rangle,\ \mathcal{C}=\{@x!=Neighborhood-2,$ $\forall t_i, t_j \in \mathcal{S}, Date(t_i)=Date(t_j) \wedge Day(t_i)=\text{"Saturday"}\}).$

3.5 The STPS Query Evaluation System

In this section we provide in depth description of the query evaluation system. We start with an overview of the indexing structures used to make the query evaluation more efficient. We then describe the *Index Join Pattern (IJP)* algorithm for evaluating pattern queries. This algorithm is based on a *merge-join* operation performed over the *inverted-indexes* corresponding to every fixed predicate in the pattern query $\mathcal{S}$.

3.5.1 Index structures

In order to efficiently evaluate pattern queries we use three indexing structures, as shown in Fig. 3.3: (a) one R-tree build on top of the BTS regions; (b) one B$^+$-tree for each BTS$_{id}$ which stores CDR records sorted by *timestamp*; and (c) one *inverted-index* for each BTS$_{id}$ which stores CDR records, sorted first by *phone$_{id}$* and then by *timestamp*, which used BTS$_{id}$ sometime during a call. Along with these indexes we also store the CDR records in the archive, grouped by *phone$_{id}$* and ordered by *timestamp*, as explained in Sect. 3.4. The R-tree is used when there is a spatio-temporal predicate in $\mathcal{S}$ which has some user defined regions (e.g. a spatial range predicate). In this case we have to find the minimal set of coverage areas from the alphabet Σ which completely cover the defined region. In order to do so, we create a range query with the user defined region and the R-tree is traversed in order to return the set of BTS that overlap with this region. The records for the returned set of BTS can be merged to form a single list with all entries to be further processed by our

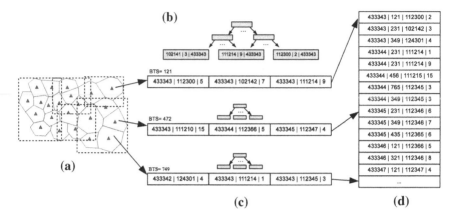

Fig. 3.3 Index framework: (**a**) R-tree for the set of BTS; (**b**) B$^+$-tree and (**c**) *Inverted-index* for each BTS; and (**d**) Original CDR archive

algorithm. This is only possible because entries in each *inverted-index* BTS$_{id}$ has its entries ordered by (*phone$_{id}$*, *timestamp*) key.

The B$^+$-tree is used by the query engine to prune entries that do not satisfy a temporal constraint. The engine makes the decision on using or not the B$^+$-tree based on the type of temporal constraint that is being evaluated (discussed later in this section). The *inverted-index* of a given BTS$_{id}$ stores pointers to all call records that are related to this BTS$_{id}$ in sometime during a call. In the *inverted-index* each entry in BTS$_{id}$ is a record that contains a *phone$_{id}$*, the *timestamp* and duration during which the user was inside region BTS$_{id}$, and a pointer to the CDR record associated to the call in the CDR archive. If a user connects to a given BTS$_{id}$ multiple times in different *timestamps*, we store a separate record for each use. Records in an *inverted-index* are ordered first by the *phone$_{id}$* and then by *timestamp*. An example of the indexing structures is shown in Fig. 3.3. The *inverted-index* entry for the region $D = 472$ is {(433342 | 124301 | 4); (433343 | 111214 | 1); (433343 | 112345 | 3); (433343 | 112345 | 20); ...}. Note that records from an *inverted-index* point to the corresponding phone user in the CDR archive. For example, the record (433343 | 111214 | 1) in the *inverted-index* 472 contains a pointer to the phone user 433343.

3.5.2 The Index-Join Pattern Algorithm (IJP)

We start with the simple scenario where the pattern S does not contain any temporal constraints. In this case, the pattern specifies only the order by which its predicates (whether fixed or variable) needs to be satisfied. Assume Q contains n predicates and let Q_f denote the subset of f fixed predicates, while Q_v denotes the subset of v variable predicates ($n = f + v$). The evaluation of Q with the proposed algorithm

Fig. 3.4 CDR examples for *inverted-indexes* $\mathcal{L}_M = 231$ and $\mathcal{L}_D = 121$

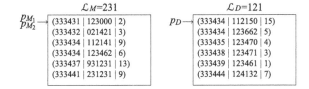

$\mathcal{L}_M = 231$

p_{M_1}, p_{M_2} →	(333431 | 123000 | 2)
	(333432 | 021421 | 3)
	(333434 | 112141 | 9)
	(333434 | 123462 | 6)
	(333437 | 931231 | 13)
	(333441 | 231231 | 9)

$\mathcal{L}_D = 121$

p_D →	(333434 | 112150 | 15)
	(333434 | 123662 | 5)
	(333435 | 123470 | 4)
	(333438 | 123471 | 3)
	(333439 | 123461 | 1)
	(333444 | 124132 | 7)

can be divided in two steps: **(i)** the algorithm evaluates the set $\mathcal{Q}_f$ using the *inverted-index* index to fast prune phone users that do not qualify for the answer; **(ii)** then the collection of the reminding *candidate* phone users is further refined by evaluating the set of variable predicates $\mathcal{S}_v$.

(i) *Fixed predicate evaluation*: All f fixed predicates in $\mathcal{Q}_f$ can be evaluated *concurrently* using an operation similar to a merge-join among their *inverted-index* lists $\mathcal{L}_i$, $i \in 1..f$. Records from these f lists are retrieved in sorted order by ($phone_{id}$, $timestamp$) and then joined by their $phone_{id}$s and $timestamp$. The join criteria is $\mathcal{L}_{i-1}.phone_{id} = \mathcal{L}_i.phone_{id}$ **and** $\mathcal{L}_{i-1}.timestamp < \mathcal{L}_i.timestamp$ (for simplicity we do not consider the dur_i attribute). The first part of the criteria ensures that we are connecting records from the same phone user and the second part ensures that we are satisfying the predicates in the appropriate order. The fact that the records in the *inverted-index* lists are sorted by ($phone_{id}$, $timestamp$) allows us to process the join with a single pass over the lists skipping all records that do not match the join criteria. If the same region appears multiple times in the pattern $\mathcal{S}$ than we use multiple pointers to the *inverted-index* lists for this region.

Example: The first step of *IJP* algorithm is illustrated using the example in Fig. 3.4. Assume the pattern $\mathcal{S}$ in the query $\mathcal{Q}$ contains three fixed and two *variable* predicates, as in: $\mathcal{S} = \{@x.M.D.@x.M\}$. This pattern looks for users that first visited some region denoted by *variable* $@x$, then visited region M sometime later (no temporal predicate is specified here), then visited region D and then visited again the same region $@x$ before finally returning to M. The first step of the join algorithm uses the *inverted-index* for M and D ($\mathcal{L}_M$ and $\mathcal{L}_D$). Conceptually, p_{M_1} and p_{M_2} represent two pointers to M *inverted-index* list.

The algorithm starts from the first record in list $\mathcal{L}_M$, phone 333431, using p_{M_1}. It then checks the first record in list $\mathcal{L}_D$, phone 333434, using p_D. We can deduce immediately that phone 333431 is not a candidate since it does not appear in the list of $\mathcal{L}_D$. So we can skip 333431 and also 333432 from the $\mathcal{L}_M$ list and continue with the next record, phone 333434. Since (333434 \| 112150 \| 15) in list $\mathcal{L}_D$ has *timestamp* greater than (333434 \| 112141 \| 9), these two occurrences of 333434 coincide with pattern $M.D$ so we need to check if 333434 uses again region M after *timestamp* 112150. Thus we consider the first record of list $\mathcal{L}_M$ using p_{M_2}, namely user (333431 \| 123000 \| 2). Since it is not from 333434 it cannot be an answer so pointer p_{M_2} advances to record (333434 \| 112141 \| 9). Now pointers in all lists point to records of 333434. However, (333434 \| 112141 \| 9) in p_{M_2} does not satisfy the pattern since its *timestamp* should be greater than *timestamp* 112150 of 333434 in D. Hence p_{M_2} is advanced to the next record, which happens to be (333434 \| 123462 \| 6). Again we

---

Algorithm 1 *IJP*: Spatial Predicate Evaluation.

Input: Query $\mathcal{S}$
Output: Phones satisfying fixed $\mathcal{S}_f$ and variable $\mathcal{S}_v$ predicates
1: Candidate Set $U \leftarrow \emptyset$, $f \leftarrow |\mathcal{S}_f|$, $Answer \leftarrow \emptyset$
2: **for** $i \leftarrow 1$ to f **do**
3:     Initialize $\mathcal{L}_i$ with the *cell-list* of $\mathcal{P}_i$
4: **for** $w \leftarrow 1$ to $|\mathcal{L}_1|$ **do**
5:     $p_1 = w$
6:     **for** $j \leftarrow 2$ to f **do**
7:         **if** $\mathcal{L}_1[w].id \notin \mathcal{L}_j$ **then break**
8:         Let k be the first entry for $\mathcal{L}_1[w].id$ in $\mathcal{L}_j$
9:         **while** $\mathcal{L}_1[w].id = \mathcal{L}_j[k].id$ **and** $\mathcal{L}_{j-1}[p_{j-1}].t > \mathcal{L}_j[k].t$ **do**
10:             $k \leftarrow k + 1$
11:         **if** $\mathcal{L}_1[w].id \neq \mathcal{L}_j[k].id$ **then break**
12:         **else** $p_j = k$
13:     **if** $\mathcal{L}_1[w]$ qualifies **then**
14:         $U \leftarrow U \cup \mathcal{L}_1[w].id$
15: **if** $|\mathcal{S}_v| = 0$ **then** $Answer \leftarrow U$
16: **else**
17:     **for** each $u \in U$ **do**
18:         $phone_{id} \leftarrow$ Retrieve(u)
19:         Build v segments S_i using $phone_{id}$
20:         Generate *variable-lists* for each segment S_i
21:         Join *variable-lists*
22:         **if** $phone_{id}$ qualifies **then**
23:             $Answer \leftarrow Answer \cup phone_{id}$

---

have a record from the same user 333434 in all lists and this occurrence of 333434 satisfies the temporal ordering, and thus the pattern $\mathcal{S}$. As a result, user 333434 is kept as a candidate in U. $\square$

In cases where a spatial predicate P_i in $\mathcal{Q}$ is a user defined area, then the above join algorithm has to materialize the *inverted-index* list for the user defined area. This materialized list has entries from the set of *inverted-index* lists for the coverage areas in the alphabet Σ which approximate the user defined area. This can be done easily since records in each *inverted-index* list in the coverage area are already ordered by ($phone_{id}$, *timestamp*). Thus, the materialized list can be computed *on-the-fly* by feeding the *IJP* algorithm with the record that has the smallest ($phone_{id}$, *timestamp*) key among the *heads* of the participating *inverted-indexes*.

(ii) Variable predicate evaluation: The second step of the *IJP* algorithm evaluates the v *variable* predicates in $\mathcal{Q}_v$, over the set of candidate phone users U generated in the first step. For a fixed predicate its corresponding *inverted-index* contains all phone users that satisfy it. However, *variable* predicates can be bound to any region, so one would have to look at all *inverted-indexes*, which is not realistic. We will again need one list for each *variable* predicate (termed *variable-list*), however such *variable-lists* are not pre-computed (like the *inverted-indexes*). Rather they are created *on-the-fly* using the candidate phone users filtered from the fixed predicate evaluation step.

Fig. 3.5 Segmentation of phone user 333434 into S_1 and S_2

S_1

(333434 | 349 | 112140 | 1)

S_2

(333434 | 125 | 123456 | 3)
(333434 | 349 | 123459 | 3)

To populate a *variable-list* for a *variable* predicate $P_i \in S_v$ we compute the possible assignments for *variable* P_i by analyzing the *inverted-index* for each candidate phone user. In particular, we use the time intervals in a candidate phone call record to identify which phone call record of the phone user can be assigned to this particular *variable* predicate. An example is shown in Fig. 3.5 using the candidate phone user 333434. From the previous step we know that 333434 satisfies the fixed predicates at the following regions: $(M, 112141)$, $(D, 112150)$, $(M, 123462)$. Using the pointers from the *inverted-indexes* of the previous step, we know where the matching regions are in the *inverted-index* of phone user 333434. As a result, the phone user 4333434 can be conceptually partitioned in two segments: phone call records that happen before $p_{M_1} = (333434 \mid 112141 \mid 9)$ are stored in S_1; and phone call records that happen after $p_D = (333434 \mid 112150 \mid 15)$ and before $p_{M_2} = (333434 \mid 123462 \mid 6)$ are stored in S_2. Note that records in between p_{M_1} and p_D do not need to be considered.

These segments are used to create the *variable-lists* by identifying the possible assignments for every *variable*. Since a *variable*'s assignments need to maintain the pattern, each *variable* is restricted by the two fixed predicates that appear before and after the *variable* in the pattern. All *variables* between two fixed predicates are first grouped together. Then for every group of *variables* the corresponding segment (the segment between two fixed predicates) is used to generate the *variable-lists* for this group. Grouping is advantageous, since it can create *variable* lists for multiple *variables* through the same pass over the phone user segments. Moreover, it ensures that the *variables* in the group maintain their order consistent with the pattern S.

Assume that a group of *variable* predicates has w members. Each segment that affects the *variables* of this group is then streamed through a window of size w. The first w elements of the phone user segment are placed in the corresponding predicate lists for the *variables*. The first element in the segment is then removed and the window shifts by one position. This proceeds until the end of the segment is reached.

The generated *variable-lists* are then joined in a way similar to the fixed predicate evaluation step. Because the *variable-lists* are populated by records coming from the same user, the join criteria checks only if the ordering of pattern S is obeyed. In addition, if the pattern contains *variables* with the same name, (in our example two @x), the join condition verifies that they are matched to the same region.

3.5.2.1 Temporal Predicate Evaluation

The *IJP* algorithm can easily support explicit temporal predicates by incorporating them as extra conditions in the join evaluations among the list records. There are three cases for a temporal predicate: (1) *interval* time $(t_{from} : t_{to})$; (2) *snapshot* time t_s; or (3) *relative* time t_r.

For the *interval* and *snapshot* temporal predicates, the B^+-tree associated to the region in the spatial predicate can be used to retrieve all phone call records that satisfy both spatial and temporal predicates. For the *interval* all records that are within the t_{from} and t_{to}, included, are retrieved, while for the *snapshot* all records that match the t_s temporal predicate are retrieved. Another approach is to verify the *interval* or *snapshot* temporal predicate for each phone call record while processing the *inverted-index* associate to a spatial predicate, without using the B^+-tree. In the next section we show that for some types of *interval* temporal predicates, evaluating the *interval* time while processing the *inverted-index* in the *IJP* algorithm is better than accessing the B^+-tree index.

For the *relative* time predicate, there are two possible strategies: (1) the straight-forward way to evaluate it is, when the spatial predicate is being evaluated, to check whether the temporal predicate is satisfied, in the same way the Algorithm 3 works; (2) another approach is to just use the B^+-tree to retrieve all records that satisfy the temporal predicate for P_i when the previous one P_{i-1} was already evaluated. The drawback of this second approach is that, every time a match for P_{i-1} occurs, a search on the B^+-tree is performed. If the number of matches for P_{i-1} is high, so the number of searches on the B^+-tree, then the first approach becomes more advantageous. Because the first approach is much simple and seems to be more efficient most of the times, we decided to always perform it when there is a *relative* temporal predicate.

3.5.2.2 Spatio-Temporal Constraints

The evaluation of spatio-temporal constraints $\mathcal{C}$ can be performed as a post filtering step after the pattern $\mathcal{S}$ evaluation is done. Other approaches to verify the set of constraints while processing the spatial predicates are also possible.

3.6 Experimental Evaluation

In this case study, we consider two real CDR databases. The first one is a CDR database from an urban environment (hereafter *Urban Database*) and the second one is a CDR database at a state level (*State Database*). The BTS hopping option was not enabled in either of the databases. Between the two databases there is no shared CDR. Furthermore, the two databases differ regarding both the number of BTSs that the infrastructure has and the spatio-temporal information available for each user (e.g., number of calls, frequency of calls, and density of BTSs). This information is to a large extent affected by the sociocultural characteristics of the regions where the data was collected from. Also, these differences deeply affect the number and characteristics of the patterns that can be detected.

Regarding the *Urban Database*, cell phone CDRs for 300,000 anonymous customers from a single carrier for a period of six months were obtained from a metropolitan area. In order to select urban users, all phone calls from a set of BTSs within the city were traced over a 2-week period (sampling period) and the (anonymous) numbers that made or received at least 3 calls per day from those BTSs were selected. Although the selection of subscribers was carried out in an urban environment, they could freely move anywhere within the nation. In total there are around 50,000,000 entries in the database considering voice, SMS and MMS. The BTS database contained the position of 30,000 towers.

As for the *State Database*, we considered 500,000 users from a state for a period of six months. No selection of users was made, i.e. all users that made or received a phone call from any BTS of that particular state during a six month period were included in the database. In total there were close to 30,000,000 entries in the database. The BTS database contained the position of 20,000 towers.

We randomly sampled 500 phone numbers from each database to generate sample queries. For each sampled phone we then randomly selected fragments in its history of calls to generate queries with varying number of predicates. Hence, these queries return at least one entry in their respective databases. For each experiment we measured the average query running time and total number of I/O for 500 queries. The query running time reports the average computational cost (as the total wall-clock time, averaged over a number of executions) for 500 queries. To maintain consistency, we set page size equals to 4 KB for indexes and data structures. All experiments were performed on a Dual Intel Xeon E5540 2.53 GHz running Linux 2.6.22 with 32 GB memory.

For evaluation purposes, we compared the *IJP* algorithm against an *extended* version of the *KMP* algorithm proposed in [21], which we call here *Extended-KMP* (*E-KMP*). The *E-KMP* supports **all** spatio-temporal features proposed in our language and it processes **all** phone users in the CDR database.

3.6.1 Comparing IJP versus KMP

In order to preserve details in all plots, we decided to suppress the *E-KMP* plots since the differences in performance between *E-KMP* and *IJP* are very large. Instead, we describe the results and comparison between both algorithms here in this section. The average number of I/O for the *E-KMP* execution is constant in both databases since it performs a sequential scanning of the phone archive. For the *State* database the number of I/O is 1,788,384 per query, while for the *Urban* it is 2,022,020. These values correspond to the total number of data disk pages each database has. The *E-KMP* algorithm performs at least 18 times more I/O than the *IJP* algorithm (for patterns with 2 user defined area predicates with a large number of BTS each for the *Urban* database). This difference is bigger if pattern queries with only spatial predicates are considered. For instance, the difference in total number of I/O for patterns with 4 spatial predicates is 108 and 260 times for the *State* and *Urban* database, respectively.

For the running time the *E-KMP* algorithm on its best performance (patterns with 4 spatial predicates for the *Urban* database) takes on average 853s per query. For the same set of experiment the *IJP* takes on average only 0.85s per query, which makes the *IJP* 1000 times faster than the *E-KMP*. Even though the cost related to I/O operations is constant when increasing the number of predicates for the *E-KMP*, the running time is not. The total time to evaluate patterns with larger number of predicates increases substantially due to the fact that more predicates have to be evaluated for a match.

3.6.2 Evaluating Patterns with Spatial Predicates

The first set of experiments evaluates patterns with different number of spatial predicates. Figure 3.6 shows the number of I/O (first row) and runtime time (second row) for 4, 8, 12 and 16 spatial predicates. For this kind of patterns only the inverted indexes associated with the predicates in the pattern are accessed. Increasing the number of spatial predicates in the query also increases the number of I/O since

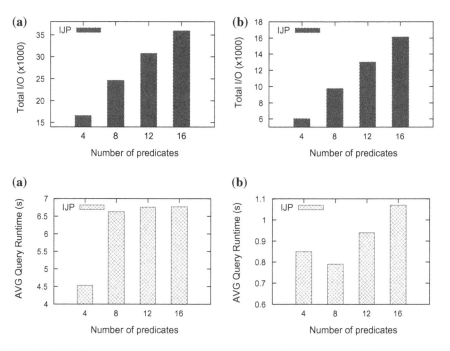

Fig. 3.6 Total I/O and query runtime for spatial predicates (**a**) State Database (**b**) Urban Database

more inverted indexes are retrieved. Also, the number of entries to be joined by the *IJP* algorithm increases, which makes the total time increase. On the average 306 and 41 phone users match a pattern for the *State* and *Urban* databases, respectively.

3.6.3 Evaluating Patterns with Variable Predicates

We also analyzed pattern queries with variables. We tested patterns with 1 variable (Fig. 3.7) and 2 variables (Fig. 3.8), varying the total number of spatial predicates from 2 to 14. For instance, in the case of patterns with 16 predicates, two query sets were generated: one with 1 variable and 15 spatial predicates; and a second one with 2 variables and 12 spatial predicates. The number of I/O for queries with 4 predicates is bigger than for queries with more predicates for some experiments. This is due to the fact that the CDR database is accessed once a match is found after the *IJP* algorithm evaluates the spatial predicates. This behavior is noticed in all the experiments except for the *Urban* database for patterns with 1 variable.

The differences in the total number of I/O for patterns with 4 predicates increase substantially from 1 to 2 variables. This is due to the fact that the number of spatial predicates drops from 3 to 2, which makes the spatial predicate evaluation phase

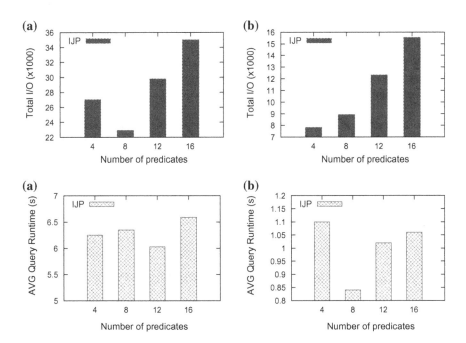

Fig. 3.7 Total I/O and query runtime for patterns with 1 variable (**a**) State Database (**b**) Urban Database

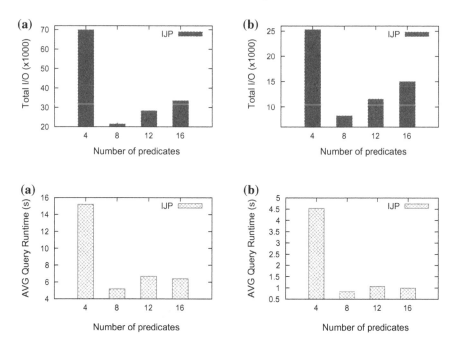

Fig. 3.8 Total I/O and query runtime for patterns with 2 variables (**a**) State Database (**b**) Urban Database

of the *IJP* algorithm less selective (there are only 2 spatial predicates to filter out CDR entries that for sure do not match the query). Therefore more CDR entries are analyzed in the variable predicate evaluation phase of the *IJP*. This behavior also occurs, but with small differences, for patterns with 8, 12 and 16 predicates. For these queries the spatial predicate evaluation phase filters out more CDR candidates than queries with only 4 predicates. Thus fewer accesses associated to the phone database are performed, reflecting in the total number of I/O shown in the plots.

   The addition of variable predicates in the pattern also increases the number of matches per query. For instance, for the *Urban* database, on average 41, 230 and 1200 phone users match for patterns with only 4 spatial predicates, 3 spatial and 1 variable predicates, and 2 spatial and 2 variable predicates respectively.

3.6.4 Evaluating Patterns with User Defined Area Predicates

In order to evaluate patterns with user defined area predicates, we generated 1 and 2 user defined area predicates by swapping a spatial predicate with an area containing a set of regions. This set of regions was selected by performing a range query on the BTS locations with center in the original spatial predicate location and a specific

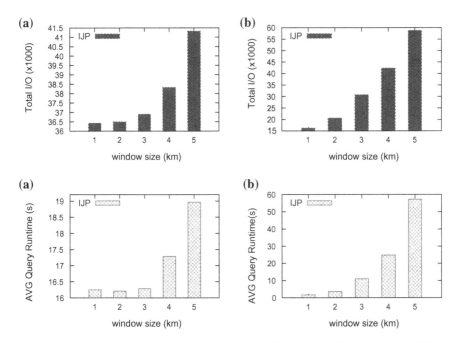

Fig. 3.9 Total I/O and query runtime for patterns with 1 defined area: (**a**) State Database (**b**) Urban Database

window size length. We then swapped the original spatial predicate with the set of regions. We generated several query sets for different window size lengths varying from 1 to $5\,km^2$. For the *Urban* database a user defined area predicate contain, on average, 2 and 400 regions for 1 and 5 km^2 window size length, respectively. For the *State* database the average number of areas selected is up to 130 regions.

Figures 3.9 and 3.10 show the results for queries with 1 and 2 user defined area predicates, respectively, for different window size lengths. For large window sizes both the total number of I/O and running time increase because more inverted indexes associated to the user defined area predicates are retrieved. Having many more entries in the inverted indexes also increases the running time since more entries are candidates to be merge-joined by the *IJP* algorithm. The same behavior is noticed when increasing the number of user defined area predicates from 1 to 2.

3.6.5 Evaluating Patterns with Temporal Predicates

In the last set of experiments we evaluated patterns with interval temporal predicates (Fig. 3.11). We generated temporal predicates from the original CDR fragments and then added them to their correspondent spatial predicate. For each pattern query all predicates have two components: a spatial and a temporal predicate. We then

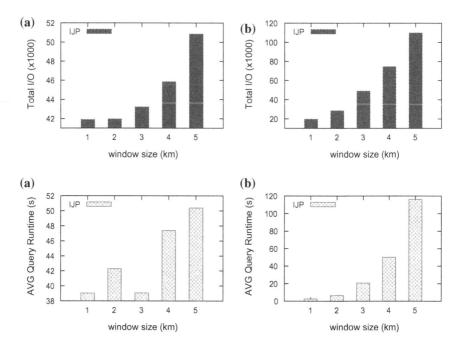

Fig. 3.10 Total I/O and query runtime for patterns with 2 defined areas (**a**) State Database (**b**) Urban Database

increased the interval in time in all temporal predicates in order to select more candidate entries. The interval values in each temporal predicate range from two to ten days covering the original timestamp of the CDR database. We evaluated patterns with temporal predicates in two ways (as explained in Sect. 3.5): the first method (*SEQ*) validates temporal predicates while processing each entry in the inverted indexes; the second method (*INDEX*) employs the B$^+$-tree, for each spatial predicate, to first evaluate the temporal predicate. In *INDEX*, entries that satisfy the temporal predicate are further grouped by $phone_{id}$ and then sorted by *timestamp* to be further processed by the *IJP* algorithm.

The total number of I/O for the *SEQ* method is constant since, for each spatial predicate, all pages in the inverted indexes are accessed. On the other hand, the number of I/O for the *INDEX* approach is much smaller than *SEQ* since only entries that satisfy the temporal predicates are retrieved. The running time of the *INDEX* approach is worse than in the *SEQ* method when increasing the interval time. This is due the factor that many more entries retrieved by the B$^+$-tree need to be further sorted before being analyzed by the *IJP* algorithm. The *INDEX* approach starts to become more appealing for temporal predicates with high selectivity (e.g. temporal predicates with interval less than 1 h (not shown in the plots)).

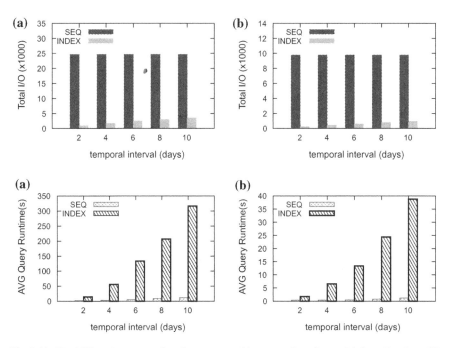

Fig. 3.11 Total I/O and query runtime for patterns with temporal predicates (**a**) State Database (**b**) Urban Database

3.7 Final Remarks

The ability to detect and characterize mobility patterns using CDR databases opens the door to a wide range of applications ranging from urban planning to crime or virus spread. Nevertheless, the spatio-temporal query systems proposed so far cannot express the flexibility that such applications require. In this case study, we described the Spatio-Temporal Pattern System (STPS) for processing spatio-temporal pattern queries over mobile phone-call databases. STPS defines a language to express pattern queries which combine fixed and variable spatial predicates with explicit and implicit temporal constraints. We described the STPS index structures and algorithm in order to efficiently process such pattern queries. The experimental evaluation shows that the STPS can answer spatio-temporal patterns very efficiently even for very large mobile phone-call databases. Among the advantages of the STPS is that it can be easily integrated in commercial telecommunication databases and also be implemented in any current commercially available RDBMS. As a future research topic, one could extend the STPS to evaluate continuous pattern queries for streaming phone-call data.

References

1. Dasgupta, K., Singh, R., Viswanathan, B., Chakraborty, D., Mukherjea, S., Nanavati, A.A., Joshi, A.: Social ties and their relevance to churn in mobile telecom networks. In: Proceedings of the International Conference on Extending Database Technology (EDBT), pp. 668–677 (2008). http://dx.doi.org/10.1145/1353343.1353424
2. Nanavati, A.A., Gurumurthy, S., Das, G., Chakraborty, D., Dasgupta, K., Mukherjea, S., Joshi, A.: On the structural properties of massive telecom call graphs: findings and implications. In: Proceedings of the ACM International Conference on Information and Knowledge Management (CIKM), pp. 435–444. ACM (2006). http://dx.doi.org/10.1145/1183614.1183678
3. Seshadri, M., Machiraju, S., Sridharan, A., Bolot, J., Faloutsos, C., Leskove, J.: Mobile call graphs: beyond power-law and lognormal distributions. In: Proceedings of the ACM SIGKDD International Conference on Knowledge Discovery and Data Mining, pp. 596–604. ACM (2008). http://dx.doi.org/10.1145/1401890.1401963
4. Gonzalez, M.C., Hidalgo, C.A., Barabasi, A.L.: Understanding individual human mobility patterns. Nature **453**, 779–782 (2008). http://dx.doi.org/10.1038/nature06958
5. Halepovic, E., Williamson, C.: Characterizing and modeling user mobility in a cellular data network. In: Proceedings of the ACM International Workshop on Performance evaluation of wireless ad hoc, sensor, and ubiquitous networks (PE-WASUN), pp. 71–78. ACM (2005). http://dx.doi.org/10.1145/1089803.1089969
6. Zang, H., Bolot, J.: Mining call and mobility data to improve paging efficiency in cellular networks. In: Proceedings of the ACM International Conference on Mobile Computing and Networking (MobiCom), pp. 123–134. ACM (2007). http://dx.doi.org/10.1145/1287853.1287868
7. Knuth, D.E., Jr., J.H.M., Pratt, V.R.: Fast pattern matching in strings. SIAM J. Comput. **6**(2), 323–350 (1977). http://dx.doi.org/10.1145/1146809.1146812
8. Sadri, R., Zaniolo, C., Zarkesh, A., Adibi, J.: Expressing and optimizing sequence queries in database systems. ACM Trans. Database Syst. **29**(2), 282–318 (2004). http://dx.doi.org/10.1145/1005566.1005568
9. Seshadri, P., Livny, M., Ramakrishnan, R.: SEQ: A model for sequence databases. In: Proceedings of the IEEE International Conference on Data Engineering (ICDE), pp. 232–239. IEEE Computer Society (1995). http://dx.doi.org/10.1109/ICDE.1995.380388
10. Agrawal, J., Diao, Y., Gyllstrom, D., Immerman, N.: Efficient pattern matching over event streams. In: Proceedings of the ACM SIGMOD International Conference on Management of Data, pp. 147–160. ACM (2008). http://dx.doi.org/10.1145/1376616.1376634
11. Erwig, M., Schneider, M.: Spatio-temporal predicates. IEEE Trans. on Knowl. and Data Eng. **14**(4), 881–901 (2002). http://dx.doi.org/10.1109/TKDE.2002.1019220
12. Mokhtar, H., Su, J., Ibarra, O.: On moving object queries. In: Proceedings of the ACM SIGMOD-SIGACT-SIGART Symposium on Principles of Database Systems (PODS), pp. 188–198. ACM (2002). http://dx.doi.org/10.1145/543613.543638
13. Hadjieleftheriou, M., Kollios, G., Bakalov, P., Tsotras, V.J.: Complex spatio-temporal pattern queries. In: Proceedings of the International Conference on Very Large Data Bases (VLDB), pp. 877–888 (2005).
14. Anagnostopoulos, A., Vlachos, M., Hadjieleftheriou, M., Keogh, E.J., Yu, P.S.: Global distance-based segmentation of trajectories. In: Proceedings of the ACM SIGKDD International Conference on Knowledge Discovery and Data Mining, pp. 34–43. ACM (2006). http://dx.doi.org/10.1145/1150402.1150411
15. Cai, Y., Ng, R.: Indexing spatio-temporal trajectories with Chebyshev polynomials. In: Proceedings of the ACM SIGMOD International Conference on Management of Data, pp. 599–610. ACM (2004). http://dx.doi.org/10.1145/1007568.1007636
16. Ni, J., Ravishankar, C.V.: PA-Tree: A parametric indexing scheme for spatio-temporal trajectories. In: Proceedings of the International Symposium on Advances in Spatial and Temporal Databases (SSTD), Lecture Notes in Computer Science, vol. 3633, pp. 254–272. Springer-Verlag Angra dos Reis, Brazil (2005). http://dx.doi.org/10.1007/11535331_15

17. Vlachos, M., Kollios, G., Gunopulos, D.: Discovering similar multidimensional trajectories. In: Proceedings of the IEEE International Conference on Data Engineering (ICDE), pp. 673–684. IEEE Computer Society (2002). http://dx.doi.org/10.1109/ICDE.2002.994784

18. Pfoser, D., Jensen, C.S., Theodoridis, Y.: Novel approaches in query processing for moving object trajectories. In: Proceedings of the International Conference on Very Large Data Bases (VLDB), pp. 395–406 (2000).

19. Hadjieleftheriou, M., Kollios, G., Tsotras, V.J., Gunopulos, D.: Indexing spatiotemporal archives. VLDB J. **15**(2), 143–164 (2006). http://dx.doi.org/10.1007/s00778-004-0151-3

20. Tao, Y., Papadias, D.: MV3R-Tree: A spatio-temporal access method for timestamp and interval queries. In: Proceedings of the International Conference on Very Large Data Bases (VLDB), pp. 431–440 (2001)

21. du Mouza, C., Rigaux, P., Scholl, M.: Efficient evaluation of parameterized pattern queries. In: Proceedings of the ACM International Conference on Information and Knowledge Management (CIKM), pp. 728–735. ACM (2005). http://dx.doi.org/10.1145/1099554.1099731

Chapter 4
Flock Pattern Queries

4.1 Introduction

Recent advances in the area of location-detection devices (e.g., RFID, GPS) and their widespread use have enabled the creation of complex tracking and situational aware-ness systems which continuously monitor the position of moving objects of interest. This huge amount of data generated by those systems motivates the need to develop efficient techniques for answering interesting queries about the past behavior of the moving objects, like discovering similarity patterns among the object trajectories.

The existing methods for querying trajectories are mainly focused on answering simple single predicate range or nearest neighbor queries (e.g., [1–4]). Examples include queries like "find all moving objects that were in area A at 10 am (in the past)" or "find the car which drove as close as possible to the location B during the time interval (10 am–1 pm)". Recently, a new group of similarity search querying methods have emerged [5–7], where the result is a trajectory closest to the query trajectory according to some metric distance (e.g., Euclidean, Dynamic Time Warping). There are also works on spatio-temporal joins (e.g. [8, 9]). Common to all of the above methods is that the query answer is validated per trajectory, i.e., a trajectory is reported to the user if its individual behavior satisfies the query predicate(s). In other words, all the above queries focus on the behavior of a trajectory as a single object and thus cannot be used to discover group patterns between the trajectories.

Lately there has also been increased interest in querying patterns capturing collab-orative or group behavior between moving objects. This includes queries like moving clusters [10, 11], convoy queries [12], swarms [13], traveling companions [14], and flocks patterns [15–17]. Such queries discover groups of moving objects that have a *strong* relationship in the space for a given time duration. The difference between all those patterns is the way they define the relationship between the moving objects and their duration in time. In this work we consider the discovery of *flock* patterns among moving objects, i.e., the problem of identifying all groups of trajectories that stay "together" for the duration of a given time interval. We consider moving objects to be "close" together if there exists a *disk* with given radius that covers all moving

M. R. Vieira and V. J. Tsotras, *Spatio-Temporal Databases*,
SpringerBriefs in Computer Science, DOI: 10.1007/978-3-319-02408-0_4,
© The Author(s) 2013

Fig. 4.1 A flock pattern
example: $\{T_1, T_2, T_3\}$

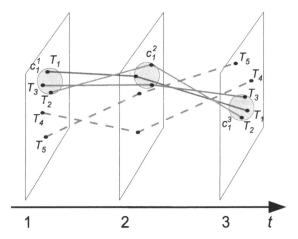

objects in the pattern (see Fig. 4.1). A trajectory satisfies the above pattern as long as
"enough" other trajectories are contained inside the disk for the specified time interval; that is, the answer is based not only on a given trajectory's behavior but also on
the trajectories near it. Such patterns are useful in security and monitoring applications, for example to potentially identify suspicious behavior within a large number
of people (e.g. "Identify all groups of five or more people that were always within a
disk of 100 feet in the last 30 min") or to study patterns of animal behavior [18–20]
(e.g., migration of sharks, whales, birds).

The example in Fig. 4.1 shows a flock pattern containing three trajectories
$\{T_1, T_2, T_3\}$ that are within a query defined disk for three consecutive time instances.
Note that the location of the disk can freely "move" in the 2-dimensional space in
order to accommodate all three moving objects and the center of the disk does not
need to coincide with any moving object location. This makes the discovery of flock
patterns difficult because there are an infinite number of possible placements of the
disk at any time instance. It is that difficulty that makes the existing methods for
flock pattern discovery [15–17] suffer from severe limitations. Such methods either
find approximate solutions or can be applied only for a single time instance of the
problem (i.e., the solution does not support the minimum time duration in the query).
To the best of our knowledge, our work is the first one to present exact solutions for
reporting flock patterns in polynomial time. It is also the first one that does so for
on-line environments. Our work is also different than clustering-based approaches
since they are not restricted to a specific shape. More details of the previous methods
are discussed in Sect. 4.2.

In this chapter, we first provide a complexity analysis for the on-line version
of the flock pattern problem. Our analysis reveals that a polynomial time solution
can be found through identifying a discrete number of locations to place the center of the flock disk inside the spatial domain. The number of such possible locations is polynomial in the total number of moving objects. Based on this analysis
we propose several evaluation algorithms that can be used to find flock patterns in

polynomial time. The first algorithm is based on time-joins, i.e., merging the results from two consecutive time instances. The other proposed methods use the *filter-and-refinement* paradigm with the purpose of reducing the total number of candidates and thus the overall computation cost of the algorithm. We evaluate our solutions using several real and synthetic moving object datasets.

The rest of the chapter is organized as follows: Sect. 4.2 highlights related work; Sect. 4.3 formally defines the on-line flock pattern and provides a complexity analysis on the problem; Sect. 4.4 describes the proposed algorithms for flock pattern discovery; Sect. 4.5 presents the performance evaluation of our proposed algorithms; Sect. 4.6 concludes this chapter with the final remarks.

4.2 Related Work

Related works can be classified in: (i) research on clustering moving objects; (ii) research on discovering convoys among trajectories; and (iii) previous works on flock discovery. Various clustering algorithms have been proposed for static spatial datasets, with different strategies ranging from partitioning (e.g. k-medoids [21]), to hierarchical (e.g., *BIRCH* [22], *CURE* [23]), and density-based (e.g., *DBSCAN* [24]). The DBSCAN algorithm works for arbitrary-shaped clusters based on the notion of density *reachability*. This method utilizes two parameters to identify dense areas: maximum distance *eps* and minimum number of points *minPts*. The DBSCAN starts with an arbitrary starting point that has not been visited. A point that has more than *minPts* within *eps* distance is considered to be in a dense area, and thus it is flagged as dense. All points inside the dense area are processed recursively the same way. Otherwise, those points are considered not *reachable* from a dense area and are labeled as outliers.

Clustering for moving objects was examined in [11], where the *DBSCAN* algorithm is performed for every time instance of the dataset. Clusters that have been found for two consecutive time instances are then joined only if the number of common objects among them is above the predefined threshold parameter θ. A cluster is reported if no other new cluster can be joined to it. This process is applied for every time instances in the dataset. Other works on clustering moving objects also include [10, 25–28]. In [10] clustering techniques were proposed to incrementally update clusters of moving objects based on the center of clusters. The object's movements are used to predict the cluster evolution over time. The *MONIC* framework [28] deals with transitions in moving clusters, e.g. disappearance and splitting. [27] presented the *microclustering* technique that groups moving objects that are not only close to each other at a specific time instance, but are also expected to move together in the near future. Recently, [25, 26] proposed to segment trajectories into line segments, then clusters are built by grouping line segments. However, time is not consider in [25, 26], which makes some line segments to be clustered together even though they are not close in space when time is considered. Nevertheless, all of the above approaches for clustering moving objects cannot solve flock pattern

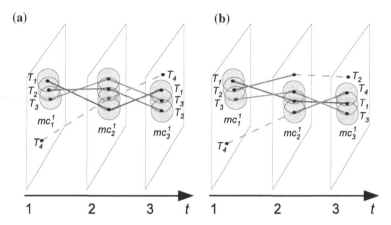

Fig. 4.2 An example comparing convoy query and moving clustering versus flock pattern query (**a**) Convoy query versus flock pattern query (**b**) Moving clustering versus flock pattern query

query since: (**1**) they use different criteria when joining the moving object clusters for two consecutive time instances; (**2**) they employ clustering algorithms, and therefore no "strong" relationship among *all* elements are enforced; and (**3**) moving clustering does not require the *same* set of moving objects to stay in a cluster *all the time* for the specified minimum duration.

Related to discovering collaborative behavior between trajectories is the work on finding *convoy patterns* in trajectory archives [12]. A convoy query is defined as a dense cluster of trajectories that stay together at least for a predefined continuous time. This type of query has four parameters: *eps* and *minPts* (the same used by the *DBSCAN* algorithm), θ' (threshold used to join clusters), and δ' (minimum duration time). *Convoy patterns* are closely related to moving clustering since both use clustering algorithms as the base of their algorithms. The main difference between these two methods is on the criteria for how clusters are joined between two consecutive timestamps. However, neither of them can solve the flock pattern query since clusters do not assume any shape restriction. For example, in Fig. 4.2a convoy query returns trajectories $\{T_1, T_2, T_3\}$ for $\theta = 3$ and for three time instances, while in Fig. 4.2b it returns nothing. For the moving clustering, if $\theta = 1$ then moving clusters return nothing in both Figs. 4.2a, b. On the other hand, if $\theta = 1/2$ then moving clustering returns $\{T_1, T_2, T_3\}$ in Fig. 4.2a and $\{T_1, T_3, T_4\}$ in Fig. 4.2b, but the last one is not a convoy query. Both examples return results based on the density of the objects, but for the flock pattern it would return nothing in either of the examples. The reason is that in both examples the objects belong to dense areas, but they do not have a *strong* interaction among other objects near them.

Flock pattern query was first introduced in [15, 16, 29] without the notion of minimum lasting time, and then reformulated later in [17] with the minimum duration as a parameter of the flock pattern. Unlike the convoy pattern, in a flock pattern the cluster has a predefined shape—a disk with radius r. A set of moving objects is considered as a flock answer if there is a disk with radius r covering the entire set of moving objects and the total number of moving objects inside the disk is greater than or equal to a given threshold. [17] shows that the discovery of the *longest* duration flock pattern is an *NP-hard* problem. As a result, [17] presents only approximation algorithms when the entire dataset is available (i.e., their approach does not work for the on-line version of the problem). To the best of our knowledge, our work is the first to propose a polynomial time solution to the on-line version of the flock pattern problem given the time duration.

4.3 Preliminaries

We assume that moving object O_{id} is uniquely identified by identifier id. Its movement is represented by a trajectory T_{id} which is defined as an ordered sequence of n multidimensional points $T_{id} = \{(l, t_1), (l, t_2), \ldots, (l, t_n)\}$. Here (l, t_i) (also represented by $l_{id}^{t_i}$) is the location of moving object O_{id} in the two dimensional space $\mathbb{R}^2$ as recorded at timestamp t_i ($t_i \in \mathbb{N}$, $t_{i-1} < t_i$, and $0 < i \leq n$). For simplicity, t_i is omitted when we discuss the current time instance, and we just use l_{id} to denote the location of moving object O_{id}.

Given two moving object locations $l_a^{t_i}$ and $l_b^{t_i}$ in a specific time instance t_i from trajectories T_a and T_b respectively, $d(l_a^{t_i}, l_b^{t_i})$ denotes the L_2 Euclidean distance between l_a and l_b. Even though here in this work we only use the L_2 distance, our methods can be generalized to other metrics as well.

A flock pattern query $Flock(\mu, \varepsilon, \delta)$ is defined as follows:

Definition 4.1 $Flock(\mu, \varepsilon, \delta)$: *Given are a set of trajectories $\mathcal{T}$, a minimum number of trajectories $\mu > 1$ ($\mu \in \mathbb{N}$), a maximum distance $\varepsilon > 0$ defined over the distance function d, and a minimum time duration $\delta > 1$ ($\delta \in \mathbb{N}$). A flock pattern $Flock(\mu, \varepsilon, \delta)$ reports all maximal size collections of trajectories $\mathcal{F}$ where: for each $f_k \in \mathcal{F}$, the number of trajectories in f_k is greater or equal than μ ($|f_k| \geq \mu$) and there exist δ consecutive time instances such that for every $t_i \in [f_k^{t_1}..f_k^{t_1+\delta}]$, there is a disk with center $c_k^{t_i}$ (center of the flock f_k at time t_i) and radius $\varepsilon/2$ covering all points in $f_k^{t_i}$. That is: $\forall f_k \in \mathcal{F}, \forall t_i \in [f_k^{t_1}..f_k^{t_1+\delta}], \forall T_j \in f_k : |f_k^{t_i}| \geq \mu, d(l_j^{t_i}, c_k^{t_i}) \leq \varepsilon/2$.*

In the above definition, a flock pattern can be viewed as a "tube" shape formed by the centers $c_k^{t_i}$, diameter ε and length δ (consecutive time instants) such that there are at least μ trajectories which stay inside the tube for the entire length δ. As an example shown in Fig. 4.3, for $Flock$ ($\mu = 3, \varepsilon, \delta = 3$) the flocks $\mathcal{F}$ reported are $f_1 = \{T_1, T_2, T_3\}$, from time instance t_1 to t_3 and disks c_1^1, c_1^2, and c_1^3, and $f_2 = \{T_4, T_5, T_6\}$, from time instance t_2 to t_4 and disks c_2^2, c_2^3, and c_2^4.

We now proceed with the complexity analysis of the flock pattern. The major challenge in this type of query is the fact that the center of the flock pattern $c_k^{t_i}$ may not belong to any of the trajectory's locations. Hence, we cannot iterate over the discrete number of all locations stored in the database and then check whether one of them

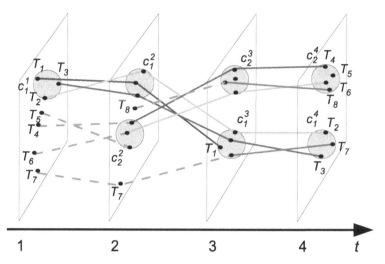

Fig. 4.3 Another example of flock pattern query with two results: $f_1 = \{T_1, T_2, T_3\}$ and $f_2 = \{T_4, T_5, T_6\}$

is a center of a flock. Moreover, since any point in the spatial domain can be a center of a flock, there are an infinite number of possible locations to test. Nevertheless, using Theorem 4.1, we show that there are a limited and discrete number of locations where we can look for flocks among the infinite number of options.

Theorem 4.1 *If for a given time instance t_i there exists a point in the space $c_k^{t_i}$ such that $\forall T_j \in f, d(l_j^{t_i}, c_k^{t_i}) \leq \varepsilon/2$ and there exists another point in the space $c'^{t_i}_k$ such that $\forall T_j \in f, d(l_j^{t_i}, c'^{t_i}_k) \leq \varepsilon/2$, then there are at least two trajectories $T_a \in f$ and $T_b \in f$ such that $\forall T_j \in \{T_a, T_b\}, d(l_j^{t_i}, c'^{t_i}_k) = \varepsilon/2$.*

Theorem 4.1 states that if there is a disk $c_k^{t_i}$ with diameter ε that covers all trajectories in the flock f at time instance t_i, then there exists another disk with the same diameter but with different center $c'^{t_i}_k$ that also covers all trajectories covered by disk $c_k^{t_i}$ and has at least two common points on its circumference. Theorem 4.1 can be easily proved by construction.

Proof Sketch. Assume that we have a disk with diameter ε and center c_k that covers all trajectories in the flock at given time instance t_i, as shown in Fig. 4.4a. For simplicity, assume that there is no trajectory point on the circumference of the disk defined by c_k and ε, i.e., $\forall T_j \in f, d(T_j, c_k) < \varepsilon/2$. We can find another disk with the same properties but with different center c'_k by using a combination of translation and rotation of the disk with center c_k. As a first step of the construction the center of the disk c_k is moved along the x axis until the first point, among all trajectories' locations, lies on the circumference of the disk. For example, in Fig. 4.4b the first point which falls on the circumference after the horizontal move of the disk center is l_1. The new center of the disk is point c'_k, and all points in the flock are covered by the

(a)   **(b)**   **(c)**

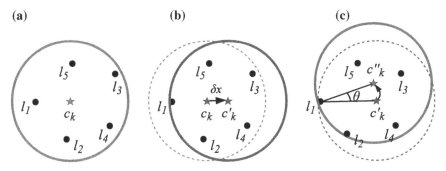

Fig. 4.4 Finding a disk $c''^{t_i}_k$ that cover a set of flock points

new disk with center c'_k and diameter ε. Otherwise, there would be a contradiction to the assumption that l_1 is the first point on the circumference. The next step of the construction rotates the new disk using as a pivot the first point l_1 on the circumference c'_k. The disk is rotated until another point falls on its circumference. In the example of Fig. 4.4c, the disk is rotated until point l_2 is on the circumference of disk c''_k. All points in the flock are still covered by the new disk with center c''_k and diameter ε (otherwise there will be a contradiction to the assumption that l_2 is the first one to be on the circumference of the disk during the rotation process). The new disk c''_k has at least two points on its circumference (points l_1 and l_2). $\square$

Theorem 4.1 has a great impact on the search for flock patterns since it limits the number of locations where to look for flocks inside the spatial domain. For a database of $|\mathcal{T}|$ trajectories there are at most $|\mathcal{T}|^2$ possible pairs of points at any given time instance. For each such pair, there are at most two disks with radius $\varepsilon/2$ that have the two points on their circumference (see Fig. 4.5). For each disk we test if it contains the required minimum number of μ trajectories. We have to perform $2|\mathcal{T}|^2$ tests for flock pattern for each time instance in the time-interval δ. The total number of possible flock patterns that need to be tested is $2|\mathcal{T}|^{2\delta}$. In order to solve the flock problem, the algorithm has to not only consider each such sequence of disks (a possible flock pattern), but also to identify the trajectories that match it. This step of checking if the trajectory stays within the sequence of disks can be done in $O(\delta)$

Fig. 4.5 Disks for $\{l_1, l_2\}$, $d(l_1, l_2) \le \varepsilon$

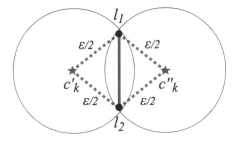

Fig. 4.6 A grid-based index
example

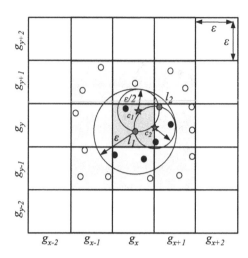

time. For the whole database it takes $O(\delta|\mathcal{T}|)$ time, and the total running time of the
algorithm will be $O(\delta|\mathcal{T}|^{(2\delta)+1})$.

4.4 Discovering Flock Patterns

In this section we describe a grid-based structure and further optimizations in order
to efficiently compute flock disks. We also describe several *on-line* algorithms to
process spatio-temporal data in an incremental fashion.

In all the proposed algorithms, we employ a grid-based structure based on grid
cells with edges of distance ε. The organization of this structure is illustrated in
Fig. 4.6. Each trajectory location $l_{id}^{t_i}$ reported for a specific time instance t_i is inserted
in a specific grid cell. The total number of cells in the index is thus affected by the
trajectory distribution in each specific time instance t_i and distance ε. The smaller
the value of ε, the larger number of grid cells are needed. Grid cells that have no
trajectory location (empty cells), are not allocated and thus do not need to be checked.
Other structures, e.g., *k-d-trees*, could be employed for organizing the trajectory's
locations. However, we chose the grid-based structure because of its simplicity, and
fast construction and query times.

Once the grid structure is built for the current time instance t_i, Algorithm 4 can
be used to find disks for t_i. For each grid cell $g_{x,y}$, only the nine adjacent grid cells,
including itself, are analyzed. Algorithm 4 analyzes points in $g_{x,y}$ and points in
$[g_{x-1,y-1}...g_{x+1,y+1}]$ in order to find pair of points (l_r, l_s) whose distances satisfy:
$d(l_r, l_s) \leq \varepsilon$. Since cells in the grid index have distance ε, for points in a particular
cell $g_{x,y}$ there is no need to analyze points further away of the range of cells in
$[g_{x-1,y-1}...g_{x+1,y+1}]$. Pairs that have not been processed yet and are within ε to
each other are further used to compute the two disks c_1 and c_2. In case that the pairs

are exactly at distance $d(l_r, l_s) = \varepsilon$, c_1 and c_2 have the same center, and thus only one needs to be analyzed.

Algorithm 1 Computing disks in grid-based index

Input: set of points $T[t_i]$ for timestamp t_i
Output: sets of *maximal* disks C
1: $C \leftarrow \emptyset$
2: $Index.Build(T[t_i], \varepsilon)$
3: **for** each non-empty cell $g_{x,y} \in Index$ **do**
4:      $L_r \leftarrow g_{x,y}$
5:      $L_s \leftarrow [g_{x-1,y-1} \cdots g_{x+1,y+1}]$
6:      **if** $|L_s| \geq \mu$ **then**
7:          **for** each $l_r \in L_r$ **do**
8:              $\mathcal{H} \leftarrow Range(l_r, \varepsilon), |\mathcal{H}| \geq \mu, d(l_r, l_s) \leq \varepsilon, l_s \in L_s$
9:              **for** each $l_j \in \mathcal{H}$ **do**
10:                  **if** $\{l_r, l_j\}$ not yet computed **then**
11:                      compute disks $\{c_1, c_2\}$ defined by $\{l_r, l_j\}$ and diameter ε
12:                      **for** each disk $c_k \in \{c_1, c_2\}$ **do**
13:                          $c \leftarrow c_k \cap \mathcal{H}$
14:                          **if** $|c| \geq \mu$ **then**
15:                              $C.Add(c)$
16:                          **end if**
17:                      **end for**
18:                  **end if**
19:              **end for**
20:          **end for**
21:      **end if**
22: **end for**

It should be noted that not all points in $[g_{x-1,y-1}...g_{x+1,y+1}]$ have to be *paired* with each point in $g_{x,y}$, but only those that have distance $d(l_r, l_s) \leq \varepsilon$ (as illustrated in Fig. 4.6). The grid-based structure can also be used to find all the points inside a given disk, as illustrated in Fig. 4.7. For each point $l_r \in g_{x,y}$ (e.g. point l_1 in Fig. 4.7a), a range query with radius ε is performed over all nine grids $[g_{x-1,y-1}...g_{x+1,y+1}]$ to find points that can be *paired* with l_r, $d(l_r, l_s) \leq \varepsilon$. Only the points in the result $\mathcal{H}$ that has at least μ trajectories ($|\mathcal{H}| \geq \mu$) are considered. For a particular disk, points in $\mathcal{H}$ are checked if they are inside the disk (Fig. 4.7b). Only the disks that have at least μ, $|c_k| \geq \mu$, points are maintained. In Fig. 4.7c disk c_1 is discarded and c_2 is considered a valid disk. Because we are interested only in *maximal* instances of flock patterns, a valid disk is further checked whether another disk has a superset of instances that the current disk has just computed. In this particular case, disks that have subset of instances are discarded and only those ones stored in C that have the *maximal* instances are returned by Algorithm 1.

The process that Algorithm 1 employs to keep only the *maximal* set of disks is based on the center of the disks and the total number of common elements between the disks. Disks are only checked with other disks that are close to each other, that is, disk c_1 is checked with c_2 only if $d(c_1, c_2) \leq \varepsilon$. If $d(c_1, c_2) > \varepsilon$, we can safely state

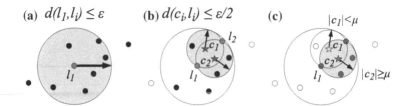

Fig. 4.7   Steps on finding flocks for time t

that they do not have any common elements. To efficiently perform the operations described above, we store disks in C using a k-d-$tree$. When checking for a particular entry c_1, we only need to check entries in the k-d-$tree$ that intersect with the new one. Only those disks that cannot be pruned are then further verified using their trajectory's identifiers. Because we store entries that belong to each disk in a binary tree, we can efficiently check if one disk has superset/subset elements of other disk. Therefore, we only need to count common elements in both disks by scanning entries in each disk. If the cardinality of common elements are $|c_1 \cap c_2| = |c_1|$ then c_1 is a subset of c_2 disk, or they have all common elements when $|c_1| = |c_2|$. Thus, c_1 can be discarded and only c_2 is kept in C. c_2 can be discarded when $|c_1 \cap c_2| = |c_2|$, otherwise both c_1 and c_2 are kept in C since one is not *maximal* with regards to the other one.

In the following subsection we describe the basic flock pattern evaluation algorithm, which evaluates flock patterns by joining the candidate disks from two consecutive time instances. We then describe four variations of the basic algorithm which use different filtering heuristics in order to reduce the number of candidate disks.

4.4.1 The Basic Flock Evaluation (BFE) Algorithm

The Basic Flock Evaluation algorithm (*BFE*) generates the candidate disks for every time instance t_i, starting with the first one t_1 and moving one time instance at a time. Each candidate disk generated in a given time instance t_i is analyzed and joined with potential flocks generated in the previous time instance t_{i-1}. Only those potential flocks that are successfully augmented with a disk in the current time instance are kept for further processing in the next time instance. This method reports flock patterns as soon as they satisfy the temporal constraint δ (i.e., we have at least δ candidate disks successfully joined in a flock).

As it was mentioned in the previous section, we use a grid-based index to find disks for the current time instance t_i. For the first time instance t_1, all disks returned by the grid-based index are stored as potential flocks (we can view a candidate disk as a partial flock with length 1) in the list of candidate flocks for this time instance $\mathcal{F}^{t_i}$. In the following time instances all disks returned by the grid-based index are stored in their candidate flock lists $\mathcal{F}^{t_i}$ and then *joined* with the candidate flocks from the previous time instance $\mathcal{F}^{t_{i-1}}$. The *join* condition used for this operation is $|c \cap f| \geq \mu$, i.e., the total number of common elements between the candidate flock and the disk has to be greater than or equal to μ. If this condition is satisfied then we move the join result into the list of candidate flocks for the current time instance t_i. A flock is found if there are at least δ join operations applied over the same candidate flock, i.e., $u.t_{end} - u.t_{start} = \delta$. In this case, the flock pattern is immediately reported to the user and its $u.t_{start}$ attribute is updated and reinserted in $\mathcal{F}^{t_i}$ to be further joined with other disks in the following time instance.

It should be noted that $\mathcal{F}^{t_i}$ only maintains potential flocks starting at some previous time instance $t_{start} > t_i - \delta$ and ending in the current time instance $t_{end} = t_i$. Entries that cannot be joined in the next time instance are discarded and not transferred into the list of candidate flocks for the next time instance.

Algorithm 2 *BFE*: Basic flock evaluation algorithm

Input: parameters μ, ε and δ
1: $\mathcal{F}^{t_0} \leftarrow \emptyset$
2: **for** each new time instance t_i **do**
3:     $\mathcal{F}^{t_i} \leftarrow \emptyset$, $C \leftarrow Index.Disks(\mathcal{T}[t_i])$
4:     **for** each $c \in C$ **do**
5:         **for** each $f \in \mathcal{F}^{t_{i-1}}$ **do**
6:             **if** $|c \cap f| \geq \mu$ **then**
7:                 $u \leftarrow c \cap f$
8:                 $u.t_{start} \leftarrow f.t_{start}$
9:                 $u.t_{end} \leftarrow t_i$
10:                **if** $(u.t_{end} - u.t_{start}) = \delta$ **then**
11:                    report flock pattern u from $u.t_{start}$ to $u.t_{end}$
12:                    update $u.t_{start}$
13:                **end if**
14:                $\mathcal{F}^{t_i} \leftarrow \mathcal{F}^{t_i} \cup u$
15:            **end if**
16:        **end for**
17:        $\mathcal{F}^{t_i} \leftarrow \mathcal{F}^{t_i} \cup c$
18:    **end for**
19: **end for**

One advantage of the *BFE* Algorithm is that for each time instance being processed, the algorithm stores only the trajectory *id*s in $\mathcal{F}^{t_i}$. There is no need to keep the actual trajectory's locations in $\mathcal{F}^{t_i}$ since they do not participate in the join condition. Another advantage is that the trajectory's locations for each time instance are processed only once, that is, there is no need to buffer trajectory data for a time window with length δ, like the other algorithms explained later in this section.

4.4.2 Filtering Heuristics

Since the number of candidate disks in a given time instance can be quite large, the total cost of joining those candidate disks can be very expensive. In order to improve the performance of the *BFE* algorithm we propose four different heuristics used to limit the number of generated candidate disks. These heuristics are described next.

4.4.2.1 The Top Down Evaluation (TDE) Algorithm

Different than the *BFE* approach that uses a *bottom-up* fashion to find flock patterns (i.e. by extending flocks one candidate disk at a time until they have length δ), the *Top Down Evaluation* (*TDE*) employs a *top-down* approach. Here we compare the candidate disks for time instances which are apart by δ. This is based on the assumption that the difference between the candidate disks in two consecutive time instances will be small (thus resulting in a large number of short flocks which still have to be kept as candidates until it becomes clear that they do not have the required length), while the differences between candidate disks from time instances which are δ time instances apart will be significantly large (and thus resulting in much fewer number of candidate flocks).

The *TDE* approach, described in Algorithm 3 uses a buffer to keep the trajectory's locations for time window w of length δ. This approach also performs a different strategy on joining the candidate disks in w. First, *BFE* calculates the candidate disks C^1 for the first time instance $t_{i-\delta+1}$ in the window w. Then, disks for the last time instance t_i in w are calculated and joined with the ones in C^1. In the last step, only the flock results that qualify from the last step are further analyzed for the remainder time instances (i.e., from $t_{i-\delta+2}$ to $t_{\delta-1}$).

4.4.2.2 The Pipe Filter Evaluation (PFE) Algorithm

The second heuristic, the *Pipe Filtering Evaluation* (*PFE*), also employs the *filter-and-refine* paradigm. This approach first filters all trajectories that have at least μ trajectories within distance ε of them for a duration of δ time instances. Then in a refinement step performed over the trajectories returned by the filtering step, it searches for flock patterns using the *BFE* Algorithm. Figure 4.8 illustrates a pipe for trajectory T_2 with radius ε. Trajectories $\{T_1, T_2, T_3, T_4\}$ are in the pipe for all δ times stamps and are further processed in the refinement step.

The *PFE* Algorithm, described in Algorithm 4 first builds a grid-based index for the first time instance $t_{i-\delta}$ in the w window. Then, it performs a range search with the center using each trajectory's location T_j in $t_{i-\delta}$. The purpose of this range search is to examine how many other trajectory's locations are within distance ε from the trajectory T_j being analyzed. If the size of the result set is greater than or equal to the threshold μ, then it applies the same procedure from time instance $t_{i-\delta+1}$ to t_i.

Algorithm 3 TDE: Top down evaluation algorithm

Input: parameters μ, ε and δ

1: **for** each new time instance t_i **do**
2:     let $\mathcal{L}$ be trajectories in windows size $|w| = \delta$ $(t_{i-\delta}...t_i)$
3:     $\mathcal{F} \leftarrow \emptyset, \mathcal{U} \leftarrow \emptyset$
4:     $C^1 \leftarrow Index.Disks(\mathcal{L}[1])$, $C^w \leftarrow Index.Disks(\mathcal{L}[w])$
5:     **for** each $c^1 \in C^1$ **do**
6:         **for** each $c^w \in C^w$ **do**
7:             **if** $|c^1 \cap c^w| \geq \mu$ **then**
8:                 $\mathcal{U} \leftarrow \mathcal{U} \cup \{c^1 \cap c^w\}$
9:             **end if**
10:         **end for**
11:     **end for**
12:     **for** each $u \in \mathcal{U}$ **do**
13:         $\mathcal{L}' \leftarrow u$, $\mathcal{F}^1 \leftarrow u^1$
14:         **for** $t \leftarrow 2$ to $|w| - 1$ **do**
15:             $\mathcal{F}^t \leftarrow \emptyset$, $C^t \leftarrow Index.Disks(\mathcal{L}'[t])$
16:             **for** each $c \in C^t$ **do**
17:                 **for** each $f \in \mathcal{F}^{t-1}$ **do**
18:                     **if** $|c \cap f| \geq \mu$ **then**
19:                         $\mathcal{F}^t \leftarrow \mathcal{F}^t \cup \{c \cap f\}$
20:                     **end if**
21:                 **end for**
22:             **end for**
23:             **if** $|\mathcal{F}^t| = 0$ **then**
24:                 **break**
25:             **end if**
26:         **end for**
27:         **for** each $f \in \mathcal{F}^{w-1}$ **do**
28:             **for** each $c^w \in C^w$ **do**
29:                 **if** $|f \cap c^w| \geq \mu$ **then**
30:                     $\mathcal{F} \leftarrow \mathcal{F} \cup \{f \cap c^w\}$
31:                 **end if**
32:             **end for**
33:         **end for**
34:     **end for**
35:     report flocks $\mathcal{F}$
36: **end for**

The resultant trajectories qualify as a candidate for a flock pattern if the total number of trajectories inside the "pipe" for a given trajectory T_j is $|\mathcal{U}| \geq \mu$. This candidate set of trajectories is further stored in the list of candidates $\mathcal{M}$ to be further processed in the refinement step of the algorithm.

In the refinement step, like in the TDE approach, the PFE Algorithm also employs the BFE Algorithm. The difference however is that now it evaluates only the trajectory's locations returned as a result of the filtering step $\mathcal{M}$, instead of using the entire trajectory database. This approach is beneficial when there is a large number of trajectories that can be pruned by the pipe filtering, and thus the procedures of

Fig. 4.8 Pipe filtering δ for T_2 and radius ε

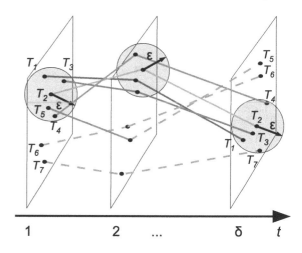

candidate disk generation and flock construction are performed over a limited subset of trajectories $m \in \mathcal{M}$.

4.4.2.3 The Continuous Refinement Evaluation (CRE) Algorithm

The *Continuous Refinement Evaluation* Algorithm (*CRE*), as the name implies, uses an heuristic that continuously refines the set of trajectories that can participate in a flock pattern. This approach uses the candidate disk generation step for time instance t_i as a filtering step for the remainder time instances t_{i+1}. That is, only trajectories that are associated with the candidate disk in time t_i are analyzed in t_{i+1}. The *CRE* approach can be used in cases where the selectivity of the candidate disks is very large, i.e., the number of candidate disks is small, as well as the total number of trajectories in them.

The pseudo code for the *CRE* algorithm is illustrated in Algorithm 8 In its first step, the *CRE* algorithm finds disks $\mathcal{C}^1$ using locations $\mathcal{L}[1]$ for time instance $t_{i-\delta}$. Then, for each disk $c^1 \in \mathcal{C}^1$ all trajectories associated with it are further processed from time instance $t_{i-\delta+1}$ to t_i. At the first time instance, disks $\mathcal{C}^1$ for time instance $t_{i-\delta}$ are stored in $\mathcal{F}^1$ (potential flocks with length 1). Then, each instance of c^1 is processed to compute disks to further be joined with disks from previous steps stored in $\mathcal{F}^t$. If $\mathcal{F}^t$ has no potential flock at time t, then the processing of c^1 can be discarded. After this second step, flock patterns are reported from time $t_{i-\delta}$ to t_i.

4.4.2.4 The Cluster Filtering Evaluation (CFE) Algorithm

The last proposed algorithm, *Cluster Filtering Evaluation* (*CFE*), has two phases: (1) the *DBSCAN* clustering algorithm with parameters *eps*=ε and *minPts*=μ is executed on the trajectory's locations for each time instance t_i; (2) then the clusters

---

Algorithm 4 *PFE*: Pipe Filter Evaluation Algorithm

Input: parameters μ, ε and δ
1: **for** each new time instance t_i **do**
2:     let $\mathcal{L}$ be trajectories in windows size $|w| = \delta$, $(t_{i-\delta}...t_i)$
3:     $\mathcal{F} \leftarrow \emptyset$
4:     **for** each $T_j \in \mathcal{L}$ **do**
5:         $\mathcal{L}' \leftarrow Index.Range(T_j, \varepsilon)$
6:         **if** $|\mathcal{L}'| \geq \mu$ **then**
7:             $\mathcal{U} \leftarrow \emptyset$
8:             **for** each $T_k \in \mathcal{L}'$ **do**
9:                 **if** $\forall t_i \in w, l_k^{t_i} \in T_k, l_j^{t_i} \in T_j, d(l_k^{t_i}, l_j^{t_i}) \leq \varepsilon$ **then**
10:                     $\mathcal{U} \leftarrow \mathcal{U} \cup T_k$
11:                 **end if**
12:             **end for**
13:             **if** $|\mathcal{U}| \geq \mu$ **then**
14:                 $\mathcal{M} \leftarrow \mathcal{M} \cup \mathcal{U}$
15:             **end if**
16:         **end if**
17:     **end for**
18:     **for** each $m \in \mathcal{M}$ **do**
19:         $\mathcal{F}^1 \leftarrow Index.Disks(m^1)$
20:         **for** $t \leftarrow 2$ to $|w|$ **do**
21:             $\mathcal{F}^t \leftarrow \emptyset$,
22:             $\mathcal{C} \leftarrow Index.Disks(m^t)$
23:             **for** each $c \in \mathcal{C}$ **do**
24:                 **for** each $f \in \mathcal{F}^{t-1}$ **do**
25:                     **if** $|c \cap f| \geq \mu$ **then**
26:                         $\mathcal{F}^t \leftarrow \mathcal{F}^t \cup \{c \cap f\}$
27:                     **end if**
28:                 **end for**
29:             **end for**
30:             **if** $|\mathcal{F}^t| = 0$ **then**
31:                 **break**
32:             **end if**
33:         **end for**
34:         $\mathcal{F} \leftarrow \mathcal{F} \cup \mathcal{F}^t$
35:     **end for**
36:     report flocks $\mathcal{F}$
37: **end for**

---

reported for a given time instance t_i are further joined with previous clusters found for t_{i-1}. The joining condition is based on the common elements between the two clusters, that is, they have to have at least μ trajectories in common. Only the clusters that satisfy the joining condition are kept. If a cluster u can be augmented in this way for δ consecutive time instances ($u.t_{end} - u.t_{start} = \delta$), then it is saved as a candidate flock, which still has to be analyzed in a refinement step using the *BFE* method. The pseudo code for the *CFE* method is summarized in Algorithm 6.

Figure 4.9 illustrates the steps performed by the *CFE* algorithm. In Fig. 4.9a, the *DBSCAN* is applied to a specific trajectory's location l_1 with parameters *eps*=ε and *minPts*=μ. Then Fig. 4.9b shows the propagation of the *DBSCAN* algorithm over l_1's

---

Algorithm 5 *CRE*: Continuous refinement evaluation algorithm

---

Input: parameters μ, ε and δ
1: **for** each new time instance t_i **do**
2:     let $\mathcal{L}$ be trajectories in windows size $|w| = \delta$, $(t_{i-\delta}...t_i)$
3:     $\mathcal{F} \leftarrow \emptyset$, $C^1 \leftarrow Index.Disks(\mathcal{L}[1])$
4:     **for** each $c^1 \in C^1$ **do**
5:         let $\mathcal{L}'$ be the trajectories in c^1 with length w
6:         $\mathcal{F}^1 \leftarrow c^1$
7:         **for** $t \leftarrow 2$ to $|w|$ **do**
8:             $\mathcal{F}^t \leftarrow \emptyset$, $C^t \leftarrow Index.Disks(\mathcal{L}'[t])$
9:             **for** each $c \in C^t$ **do**
10:                 **for** each $f \in \mathcal{F}^{t-1}$ **do**
11:                     **if** $|c \cap f| \geq \mu$ **then**
12:                         $\mathcal{F}^t \leftarrow \mathcal{F}^t \cup \{c \cap f\}$
13:                     **end if**
14:                 **end for**
15:             **end for**
16:             **if** $|\mathcal{F}^t| = 0$ **then**
17:                 **break**
18:             **end if**
19:         **end for**
20:         $\mathcal{F} \leftarrow \mathcal{F} \cup \mathcal{F}^t$
21:     **end for**
22:     report flocks $\mathcal{F}$
23: **end for**

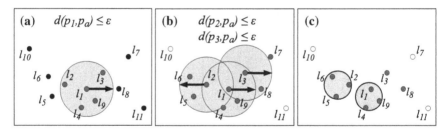

Fig. 4.9 *CFE* steps to find flock patterns

neighbors. Trajectory's locations that do not belong to any cluster are then discarded. In Fig. 4.9c, the final two clusters ($\{l_2, l_5, l_6\}$ and $\{l_1, l_4, l_9\}$) reported by the *DBSCAN* algorithm are then further analyzed in the refinement step by the *CFE* Algorithm.

4.5 Experimental Evaluation

In order to evaluate the performance of the proposed methods, we performed several experiments with several trajectory datasets and different parameters. In particular we show the results for one (*SG*) synthetic and four real (*Trucks*, *Buses*, *Cars*

Algorithm 6 CFE: Clustering filtering evaluation algorithm

Input: parameters μ, ε and δ
1: $\mathcal{I}^{t_i} \leftarrow \emptyset$
2: **for** each new time instance t_i **do**
3:     $\mathcal{U} \leftarrow \emptyset$, $\mathcal{L} \leftarrow T[t_i]$
4:     $\mathcal{Q} \leftarrow DBSCAN(\mathcal{L}, \mu, \varepsilon)$
5:     **for** each $q \in \mathcal{Q}$ **do**
6:         **for** each $f \in \mathcal{I}^{t_{i-1}}$ **do**
7:             **if** $|q \cap f| \geq \mu$ **then**
8:                 $u \leftarrow \{q \cap f\}$
9:                 $u.t_{start} \leftarrow f.t_{start}$
10:                 $u.t_{end} \leftarrow t$
11:                 **if** $(u.t_{end} - u.t_{start}) = \delta$ **then**
12:                     $\mathcal{F}^1 \leftarrow Index.Disks(u^1)$
13:                     **for** $t \leftarrow 1$ to $|w|$ **do**
14:                         $\mathcal{F}^t \leftarrow \emptyset$, $C \leftarrow Index.Disks(m^t)$
15:                         **for** each $c \in C$ **do**
16:                             **for** each $f \in \mathcal{F}^{t-1}$ **do**
17:                                 **if** $|c \cap f| \geq \mu$ **then**
18:                                     $\mathcal{F}^t \leftarrow \mathcal{F}^t \cup \{c \cap f\}$
19:                                 **end if**
20:                             **end for**
21:                         **end for**
22:                         **if** $|\mathcal{F}^t| = 0$ **then**
23:                             **break**
24:                         **end if**
25:                     **end for**
26:                     $\mathcal{F} \leftarrow \mathcal{F} \cup \mathcal{F}^t$
27:                     update $u.t_{start}$
28:                 **end if**
29:                 $\mathcal{U} \leftarrow \mathcal{U} \cup u$
30:             **end if**
31:         **end for**
32:         $\mathcal{U} \leftarrow \mathcal{U} \cup q$
33:     **end for**
34:     $\mathcal{I}^{t_i} \leftarrow U$
35: **end for**

---

and *Caribous*) datasets. The first two real datasets, *Trucks* and *Buses* [30], contain 112,203 and 66,096 moving object locations generated from 276 and 145 moving trucks and buses, respectively, collected in the greater metropolitan area of Athens, Greece. The third dataset *Cars* [31] contains 134,263 object locations collected from 183 private cars moving in Copenhagen, Denmark. The last real trajectory dataset *Caribous* [18] is generated from the analysis of the migration movements of 43 caribous in northwestern states of Canada. The size of the dataset is 15,796 object locations.

Since the real datasets that we could find in public domain are relatively small, we also use a synthetic dataset *SG* in order to test the scalability of our methods. The *SG* dataset is generated by simulating the movement of 50,000 vehicles on

Table 4.1 Parameters values for each dataset

Dataset	μ [default] min #traj.	ε [default] max dist.	δ [default] min time
Trucks	4, 6,..., 20 **[5]**	0.8, 0.9, ..., 1.5 **[1.2]**	4, 6, ..., 20 **[10]**
Cars	4, 6, ..., 20 **[5]**	0.8, 0.9, ..., 1.5 **[1.2]**	4, 6, ..., 20 **[10]**
Caribous	2, 3, ..., 10 **[5]**	0.1, 0.2, ..., 0.8 **[1.6]**	4, 6, ..., 20 **[10]**
Buses	4, 6, ..., 20 **[5]**	0.4, 0.5, ..., 1.1 **[1.2]**	4, 6, ..., 20 **[10]**
SG	4, 6, ..., 20 **[5]**	2.2, 2.6, ..., 5.0 **[3.4]**	4, 6, ..., 20 **[10]**

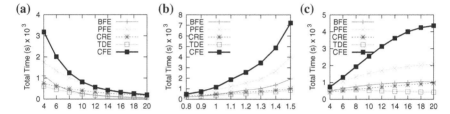

Fig. 4.10 Total time for the *Trucks* dataset (**a**) varying μ (*min* #traj.) (**b**) varying ε (*max dist.*) (**c**) varying δ (*min* time)

road network of Singapore. Those moving objects have different velocities and their starting locations were randomly placed in the road network. The size of the synthetic dataset is 2,548,084 moving object locations.

In our experiments we use several values for the flock parameters μ, ε and δ. The ranges of values for each dataset are shown in Table 4.1, where in bold we show the default values for each parameter. Those default values are used when the value of the parameter is not explicitly specified for a given experimental set. The total numbers of patterns discovered for the minimum and maximum value of each parameter, taken from Table 4.1, are shown in Table 4.2.

Figures 4.10, 4.11, 4.12, 4.13, 4.14 show the results when varying the parameters μ (first column), ε (second column) and δ (third column) for the five different datasets. All plots show the total time (in seconds) needed to process the whole dataset, including building the grid index. As it can be seen, when increasing μ, decreasing ε, or decreasing δ, the total time needed to discover the flock patterns for all the proposed methods decreases. This is expected since the flock queries become more selective, and thus all the algorithms have to maintain fewer candidate trajectories during the evaluation.

For the *Trucks* and *Cars* datasets, the *TDE* and *CRE* methods have significantly better performance compared to the other three methods. The gap in performance between these two methods and the other ones increases when the selectivity of the queries becomes low (for small μ and large ε). This is due to the fact that the large number of partial intermediate results that have to be maintained and processed by the *BFE*, *PFE* and *CFE* methods. Similar behavior can be observed for large values

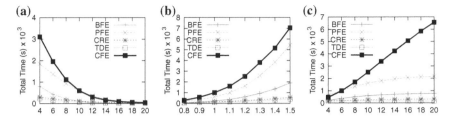

Fig. 4.11 Total time for the *Cars* dataset (**a**) varying μ (*min* #traj.) (**b**) varying ε (*max* dist.) (**c**) varying δ (*min* time)

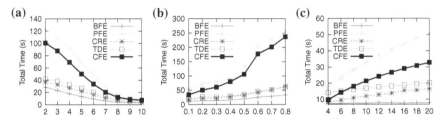

Fig. 4.12 Total time for the *Caribous* dataset (**a**) varying μ (*min* #traj.) (**b**) varying ε (*max* dist.) (**c**) varying δ (*min* time)

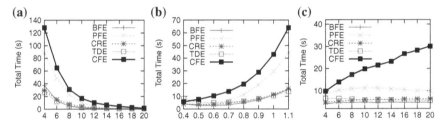

Fig. 4.13 Total time for the *Buses* dataset (**a**) varying μ (*min* #traj.) (**b**) varying ε (*max* dist.) (**c**) varying δ (*min* time)

of δ, but only for the *PFE* and *CFE* methods. This is due to the fact that these two methods keep the trajectory history in a time window w before computing the disks for each timestamp. Similar behaviors are observed for the *Buses* dataset.

For the *Caribous* dataset the *BFE* algorithm achieved the best performance, closely followed by the *TDE* and *CRE*. Our analysis shows that the *BFE* algorithm performed well in this dataset because of the pattern of the trajectories' movements. The movements of the moving objects in this dataset seem to be very well correlated, e.g., all 43 caribous have similar migration patterns and stay very close together (i.e., they are grouped in herds during their movement). Because of this, the other methods are not able to prune a lot of trajectories in their filtering phases. The fact that the data in the *Caribous* dataset follows the pattern of the flock queries can also be observed by the total number of flocks discovered for this dataset (see Table 4.2).

Table 4.2 Number of flock patterns discovered

Dataset	μ - min #traj.		ε - max dist.		δ - min time	
	min	max	min	max	min	max
Trucks	309	14,935	3,741	15,608	2,045	23,222
Cars	62	18,451	3,218	23,440	3,149	24,211
Caribous	124	9,480	5,292	6,915	3,364	4,598
Buses	0	2,988	16	1,021	55	1,730
SG	0	1,304	53	741	112	385

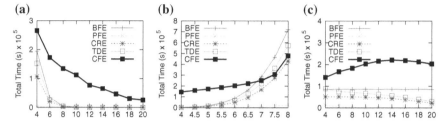

Fig. 4.14 Total time for the *SG* dataset (**a**) varying μ (*min* #traj.) (**b**) varying ε (*max* dist.) (**c**) varying δ (*min* time)

The number of discovered flocks is quite large for a dataset with only 15,796 moving object locations.

In the next group of experiments, we test the performance of the proposed methods using the synthetic dataset *SG*. As it can be seen from the plots, the *PFE* algorithm is by far the best algorithm for this particular dataset. The main reason for this behavior is that even though the total number of potential flocks for each timestamp can be quite big (see Table 4.3 for details), the *PFE* approach performs more holistic filtering compared to the other solutions. This heuristic is checking the minimum occupancy criteria (the number of trajectories closer than the threshold ε to a given trajectory should be more than μ) for the *entire* duration δ of the flock pattern query. As for the other four methods, they all join candidate disks for two consecutive timestamps without considering the holistic view of the trajectories' movements. Therefore, the first filtering phase of the *PFE* has a higher pruning capability compared to the other methods for the *SG* dataset. We should note that in the real datasets, trajectories follow similar patterns, while in the *SG* dataset objects follow random patterns and though they might be close together in one time instance, they tend not to follow similar patterns for several consecutive timestamps.

As it can be seen from the plots, for most of the datasets the *CFE* algorithm has the worst performance among the proposed methods. This is due to the fact that the filtering step in the *CFE* approach employs a clustering technique, which can be very expensive for large datasets. This approach however works significantly better when the datasets are relatively small and the trajectory's movements follow similar moving patterns (see the results for the *Caribous* dataset). In such cases, the high

Table 4.3 Min/Max Number of disks per time

Dataset	μ - min #traj.		ϵ - max dist.		δ - min time	
	min	max	min	max	min	max
Trucks	505	1,257	812	1,547	1,237	1,237
Cars	72	294	142	387	279	279
Caribous	393	235	587	342	309	309
Buses	7	236	27	183	105	105
SG	1,343	12,894	1,232	2,916	10,934	10,934

computational cost for the clustering phase pays off when considered the final cost of the *CFE* algorithm.

In our next set of experiments, we measure the minimum and the maximum number of disks computed for each time instance using our grid-based index. The results can be depicted in Table 4.3. As it can be seen even for big values of the parameters μ, ε and δ, the maximum number of disks computed per timestamp is relatively small compared with the number of trajectories. This shows the efficiency of our grid-based index structure.

4.6 Final Remarks

Recently there has been increased interest in queries that capture the collaborative behavior of spatio-temporal data (e.g., convoys, flocks). In particular, a flock pattern contains a group of at least μ moving objects which are all "enclosed" by a disk of diameter ε for at least δ consecutive timestamps. Discovering on-line flock patterns is useful for several applications, ranging from tracking suspicious activities to migrations of animals. Previous related works either do not work for on-line datasets, or do not return exact solutions for flock pattern queries. In this chapter, we first showed that the discovery of flock patterns with fixed time duration can be computed in polynomial time. We then presented a framework that uses a lightweight grid-based structure in order to efficiently and incrementally process the trajectory's locations. Using this framework, we proposed various on-line flock discovery algorithms. Experiments on various trajectory datasets showed that the proposed methods can efficiently report flock patterns even for large datasets and for different variations of the flock parameters (μ, ε and δ). As future work, one could examine cost models to enable the user pick the most efficient algorithm based on the data distribution and query parameters.

References

1. Aggarwal, C.C., Agrawal, D.: On nearest neighbor indexing of nonlinear trajectories. In: Proceedings of the ACM SIGMOD-SIGACT-SIGART Symposium on Principles of Database Systems (PODS), pp. 252–259. ACM (2003). DOI http://dx.doi.org/10.1145/773153.773178
2. Benetis, R., Jensen, S., Karciauskas, G., Saltenis, S.: Nearest and reverse nearest neighbor queries for moving objects. The VLDB Journal 15(3), 229–249 (2006). DOI http://dx.doi.org/10.1007/s00778-005-0166-4
3. Pfoser, D., Jensen, C.S., Theodoridis, Y.: Novel approaches in query processing for moving object trajectories. In: Proceedings of the International Conference on Very Large Data Bases (VLDB), pp. 395–406 (2000)
4. Tao, Y., Papadias, D., Shen, Q.: Continuous nearest neighbor search. In: Proceedings of the International Conference on Very Large Data Bases (VLDB), pp. 287–298 (2002)
5. Cai, Y., Ng, R.: Indexing spatio-temporal trajectories with Chebyshev polynomials. In: Proceedings of the ACM SIGMOD International Conference on Management of Data, pp. 599–610. ACM (2004). DOI http://dx.doi.org/10.1145/1007568.1007636
6. Lee, S.L., Chun, S.J., Kim, D.H., Lee, J.H., Chung, C.W.: Similarity search for multidimensional data sequences. In: Proceedings of the IEEE International Conference on Data Engineering (ICDE), pp. 599–608. IEEE Computer Society (2000). DOI http://dx.doi.org/10.1109/ICDE.2000.839473
7. Vlachos, M., Kollios, G., Gunopulos, D.: Discovering similar multidimensional trajectories. In: Proceedings of the IEEE International Conference on Data Engineering (ICDE), pp. 673–684. IEEE Computer Society (2002). DOI http://dx.doi.org/10.1109/ICDE.2002.994784
8. Arumugam, S., Jermaine, C.: Closest-point-of-approach join for moving object histories. In: Proceedings of the IEEE International Conference on Data Engineering (ICDE), pp. 86–86. IEEE Computer Society (2006). DOI http://dx.doi.org/10.1109/ICDE.2006.36
9. Bakalov, P., Hadjieleftheriou, M., Tsotras, V.J.: Time relaxed spatiotemporal trajectory joins. In: Proceedings of the ACM SIGSPATIAL International Conference on Advances in Geographic Information Systems, pp. 182–191. ACM (2005). DOI http://dx.doi.org/10.1145/1097064.1097091
10. Jensen, C.S., Lin, D., Ooi, B.C.: Continuous clustering of moving objects. IEEE Trans. on Knowl. and Data Eng. 19(9), 1161–1174 (2007). DOI http://dx.doi.org/10.1109/TKDE.2007.1054
11. Kalnis, P., Mamoulis, N., Bakiras, S.: On discovering moving clusters in spatio-temporal data. In: Proceedings of the International Symposium on Advances in Spatial and Temporal Databases (SSTD), *Lecture Notes in Computer Science*, vol. 3633, pp. 364–381. Springer (2005). DOI http://dx.doi.org/10.1007/11535331_21
12. Jeung, H., Yiu, M.L., Zhou, X., Jensen, C.S., Shen, H.T.: Discovery of convoys in trajectory databases. Proceedings of the VLDB Endowment (PVLDB) 1(1), 1068–1080 (2008)
13. Li, Z., Ding, B., Han, J., Kays, R.: Swarm: Mining relaxed temporal moving object clusters. Proceedings of the VLDB Endowment (PVLDB) 3(1), 723–734 (2010)
14. Tang, L.A., Zheng, Y., Yuan, J., Han, J., Leung, A., Hung, C.C., Peng, W.C.: On discovery of traveling companions from streaming trajectories. In: Proceedings of the IEEE International Conference on Data Engineering (ICDE), pp. 186–197. IEEE Computer Society (2012). DOI http://dx.doi.org/10.1109/ICDE.2012.33
15. Benkert, M., Gudmundsson, J., Hübner, F., Wolle, T.: Reporting flock patterns. In: Proceedings of the Conference on Annual European Symposium on Algorithms (ESA), pp. 660–671. Springer-Verlag (2006). DOI http://dx.doi.org/10.1007/11841036_59
16. Benkert, M., Gudmundsson, J., Hübner, F., Wolle, T.: Reporting flock patterns. Comput. Geom. Theory Appl. 41, 111–125 (2008). DOI http://dx.doi.org/10.1016/j.comgeo.2007.10.003

17. Gudmundsson, J., van Kreveld, M.: Computing longest duration flocks in trajectory data. In: Proceedings of the ACM SIGSPATIAL International Conference on Advances in Geographic Information Systems, pp. 35–42. ACM (2006). DOI http://dx.doi.org/10.1145/1183471.1183479

18. Board, P.C.M.: Porcupine caribou herd satellite collar project. www.taiga.net/satellite (2009)

19. Environmental Studies: www.environmental-studies.de (2009)

20. WhaleNet: http://whale.wheelock.edu (2010)

21. Ng, R.T., Han, J.: Efficient and effective clustering methods for spatial data mining. In: Proceedings of the International Conference on Very Large Data Bases (VLDB), pp. 144–155 (1994)

22. Zhang, T., Ramakrishnan, R., Livny, M.: BIRCH: An efficient data clustering method for very large databases. In: Proceedings of the ACM SIGMOD International Conference on Management of Data, pp. 103–114. ACM (1996). DOI http://dx.doi.org/10.1145/235968.233324

23. Guha, S., Rastogi, R., Shim, K.: Cure: An efficient clustering algorithm for large databases. In: Proceedings of the ACM SIGMOD International Conference on Management of Data, pp. 73–84. ACM (1998). DOI http://dx.doi.org/10.1145/276304.276312

24. Ester, M., Kriegel, H.P., Sander, J., Xu, X.: A density-based algorithm for discovering clusters in large spatial databases with noise. In: E. Simoudis, J. Han, U.M. Fayyad (eds.) KDD, pp. 226–231. AAAI Press (1996).

25. Lee, J.G., Han, J., Li, X., Gonzalez, H.: TraClass: trajectory classification using hierarchical region-based and trajectory-based clustering. Proceedings of the VLDB Endowment (PVLDB) **1**, 1081–1094 (2008)

26. Lee, J.G., Han, J., Whang, K.Y.: Trajectory clustering: a partition-and-group framework. In: Proceedings of the ACM SIGMOD International Conference on Management of Data, pp. 593–604. ACM (2007). DOI http://dx.doi.org/10.1145/1247480.1247546

27. Li, Y., Han, J., Yang, J.: Clustering moving objects. In: Proceedings of the ACM SIGKDD International Conference on Knowledge Discovery and Data Mining, pp. 617–622. ACM (2004). DOI http://dx.doi.org/10.1145/1014052.1014129

28. Spiliopoulou, M., Ntoutsi, I., Theodoridis, Y., Schult, R.: Monic: modeling and monitoring cluster transitions. In: Proceedings of the ACM SIGKDD International Conference on Knowledge Discovery and Data Mining, pp. 706–711. ACM (2006). DOI http://dx.doi.org/10.1145/1150402.1150491

29. Laube, P., Kreveld, M., Imfeld, S.: Finding remo âĂT detecting relative motion patterns in geospatial lifelines. In: Developments in Spatial Data Handling, pp. 201–215. Springer, Berlin Heidelberg (2005). DOI http://dx.doi.org/10.1007/3-540-26772-7_16

30. The Chorochronos Archive: The R Tree Portal. http://www.chorochronos.org (2013)

31. Jensen, C.S.: Daisy. http://www.daisy.aau.dk (2011)

Chapter 5
Diversified Pattern Queries

5.1 Introduction

Many database and information retrieval applications have recently started to incorporate capabilities to rank elements with respect to *relevance* and *diversity* features, i.e., the retrieved elements should be as relevant as possible to the query, and, at the same time, the result set should be as diverse as possible. Examples of such applications range from exploratory and ambiguous keywords searches (e.g., jaguar, apple, java, windows, eclipse) [5, 12, 29], diversification of structured databases [8, 24] and user personalized results [27], to topic summarization [2, 6, 23], or even to exclude near-duplicate results from multiple resources [15]. While addressing relevance is comparatively straightforward, and has been heavily studied in both database and information retrieval areas, diversity is a more difficult problem to solve, where optimal evaluations have worst-case NP-hard computation time [1, 3].

Typically, in all of the above applications, the final result set is computed in two phases. First, a ranking *candidate set S* is retrieved with elements that are relevant to the user's query. Then, in the second phase, a *result set R*, $R \subseteq S$, is computed containing very *relevant* elements to the query and, at the same time, as *diverse* as possible to other elements in the result set R. Since these two components relevance and diversity *compete* with each other, previous algorithms for query result diversification attempt to find a *tradeoff* between the *relevance* and *diversity* components. Thus, the query result diversification problem can be modeled as a bi-criteria optimization problem. One advantage of using the *tradeoff* parameter to tune the importance between *relevance* and *diversity* is that the user can give more preference on one of these two components. For instance, if a candidate set has a large amount of near-duplicate elements, then a user can increase the *tradeoff* value, as necessary, in order to have more diverse elements in the result set.

Several techniques have been introduced for diversifying query results, with the majority of them exploring a greedy solution that builds the result set in an incremental way [2, 7, 16, 34, 36, 37]. These techniques typically choose the first element to be added to the result set based only on relevance; further elements are added

M. R. Vieira and V. J. Tsotras, *Spatio-Temporal Databases*,
SpringerBriefs in Computer Science, DOI: 10.1007/978-3-319-02408-0_5,
© The Author(s) 2013

based on an element's relevance and diversity against the current result set. The basic assumption of these techniques is that the result set does not change with the size k of the result set, i.e., $R \subset R'$, $|R| = k$ and $|R'| = k + 1$, which is not always true. In some of the above techniques, there is the additional problem that they are not able to handle different values of *tradeoff* between relevance and diversity; and for the few ones that do support it, e.g., [2, 12], as we show in our experiments they do not perform well.

In this chapter we describe the DivDB framework, which uses a bi-criteria optimization objective function similar to [12] to compare and evaluate different methods of diversifying query results. Gollapudi and Sharma [12] proposed a set of natural axioms that a diversification system is expected to satisfy and showed that no diversification objective can satisfy all the axioms simultaneously. Since there is no single objective function that is suitable for every application domain, the authors describe three diversification objective functions: *max-sum diversification*, *max-min diversification* and *mono-objective*. In our work we particularly focus on the *max-sum diversification* since it can be reduced to different versions of the well-studied facility dispersion problem [19, 26], for which efficient approximation algorithms exist (e.g., [10, 14, 17, 18, 33]). Nevertheless, our work can be easily extended to other functions, e.g. *max-min, max-avg*.

Since the methods described here and implemented in the DivDB framework rely only on the relevance and diversity values, our work can be employed in *any* domain where relevance and diversity functions can be defined. Moreover, the evaluation of the different methods described in this chapter is done using only the values computed by the optimization objective function, and not using any external information (e.g., subtopic coverage).

In addition, we also describe two new methods that incrementally construct the result set using a scoring function. This function ranks candidate elements using not only relevance and diversity to the existing result set, but also accounts for diversity against the remaining candidate elements. We present a thorough experimental evaluation of several of the diversification techniques implemented in the DivDB framework. Our experimental results show that the two proposed methods achieve precision very close to the optimal, while also providing the best result quality measured using the optimization objective function.

To summarize, in this work we make the following contributions:

1. we present the DivDB framework for evaluation and optimization of methods for diversifying query results, where a user can adjust the *tradeoff* between relevance and diversity;
2. we adapt and evaluate several existing methods using the DivDB framework;
3. we propose a new function to compute the contribution of each candidate element to the result set. This function not only considers the relevance value between the candidate element and the query, but also how diverse it is to other elements in the candidate and result sets;
4. we propose two new methods for diversifying query results using the above contribution function;

5. we perform a thorough experimental evaluation of the various existing diversification techniques.

The rest of the chapter is organized as follows: Section 5.2 describes the related
work on query result diversification and *Max-Sum Dispersion Problem*; Sect. 5.3
defines the query result diversification problem and the DivDB framework used for
evaluation and optimization of methods for diversifying query results; Sects. 5.4 and
5.5 describe, respectively, previously known and two new methods using the DivDB
framework; the experimental evaluation is shown in Sects. 5.6 and 5.7 concludes this
chapter with the final remarks.

5.2 Related Work

We first present related work on query result diversification and then on the *Max-Sum
Dispersion Problem*, which relates to our GNE method.

5.2.1 Query Result Diversification

Result diversification has recently been examined and applied to several different
domains [13]. In [21] clustering techniques are employed in order to generate diverse
results, while in [28] learning algorithms based on users' clicking behavior are used to
re-rank elements in the result set. Techniques to extract compact information from
retrieved documents in order to test element's coverage with respect to the query
(and at the same time, avoid excessive redundancy among elements in the result) are
explored in [6, 23]. Similar techniques are employed for structured query results in
[8, 24]. Topic diversification methods based on personalized lists in recommendation
systems are proposed in [42]. In [27, 31, 32] techniques to generate related queries
to the user's query are employed to yield a more diverse result set for documents.
Zhai et al. [38] proposes a risk minimization framework where users can define a
loss function over the result set which leads to a diverse result.

Greedy algorithms for explanation-based diversification for recommended items
are proposed in [37]. Unlike ours, this approach considers the relevance and diversity
of explanation items as two separate components. Therefore, to optimize both relevant and diverse explanations, the proposed algorithms rely on parameters' values
that typically are not easy to tune.

Agrawal et al. [1] describe a greedy algorithm which relies on an objective function
computed based on a probabilistic model that admits a sub-modularity structure
and a taxonomy to compute diverse results. Relevance of documents to queries is
computed using standard ranking techniques, while diversity is calculated through
categorization according to the taxonomy. The aim is to find a set of documents that
covers several taxonomies of the query, that is, the result set is diverse for the defined

query using the taxonomy. In [36], it is presented two greedy algorithms to compute a result set; here the elements in the result set are diverse regarding their frequency in the data collection. However, the proposed algorithms work only on structural data with explicitly defined relationships. In [5], a Bayesian network model is used as a blind negative relevance feedback to *re-rank* documents, assuming that in each position in the rank order all previous documents are irrelevant to the query. An affinity graph is used in [39] to penalize redundancy by lowering an item in the rank order if already seen elements appeared in the rank order. In [41], absorbing Markov chain random walks is used to *re-rank* elements, where an item that has already been ranked becomes an absorbing state, dragging down the importance of similar unranked states.

The main difference between all these works and the ones described in this chapter is that we do not rely on external information to generate diverse results, e.g., taxonomy, known structure of the data, click-through rates, and query logs. Since workloads and queries are rarely known in advance, the external information discussed above is typically expensive to compute and provides suboptimal results. Instead, the methods presented here rely only on computed values using relevance (similarity) and diversity functions (e.g., tf/idf cosine similarity, Euclidean distance) in the data domain.

The Maximal Marginal Relevance (MMR) [2] is one of the earliest proposals to use a function that *re-ranks* elements in the result set in order to generate relevant elements to the query, and, at the same time, diverse to the previously selected elements in the result. Variations of MMR were also proposed for different domains [4, 7, 9]. At each iteration, the MMR function returns the element with the highest value with respect to a *tradeoff* between relevance and diversity to only the current result set. Our proposed GMC and GNE methods also use a function that computes a *tradeoff* between relevance and diversity. However, they both use a function that computes the maximum contribution that an element can provide to the **final result**. Moreover MMR always picks as first the element that is most relevant to the query, which can highly affect the other elements that are chosen in the result. The MMR is one of several methods that we implemented and compared against other ones in the DivDB framework.

In [16] a greedy strategy is proposed to compute the result set based on a *Boolean* function: one element is included in the result set if it has diversity value more than a predefined threshold to all elements in the current result set. Thus, the diversity function is used only to filter out elements in the result set. The aim is to maximize the relevance while the result set's diversity remains above a certain threshold. This is different from our methods which maximize the objective function with respect to relevance **and** diversity. Furthermore, the threshold parameter has a great influence on the performance of the algorithm, which can easily affect the quality of the results.

5.2.2 Max-Sum Dispersion Problem

Gollapudi and Sharma [12] proposed a generic framework with eight axioms that a diversification system is expected to satisfy. The problem formulation that most relates to our work is the max-sum diversification. Gollapudi and Sharma [12] also shows how to reduce max-sum diversification to a max-sum dispersion problem (also known as the p-dispersion in the Operations Research literature [19, 26]). Given that the objective function is a metric, it then uses an approximation algorithm [14] to solve the problem.

Instead of an approximation approach, our second proposed method GNE uses a randomized algorithm to solve the max-sum diversification problem. In particular, we use an adaptation of a meta-heuristic called Greedy Randomized Adaptive Search Procedure (GRASP) proposed in [10, 20, 25, 30, 33]. To the best of our knowledge, our proposal is the first randomized approach applied to the result diversification problem. As it is shown in the experimental section, the approximation algorithm in [12] is very expensive to compute and provides results with limited precision when compared to the optimal solution.

GRASP is a greedy randomized method that employs a multistart procedure for finding approximate solutions to combinatorial optimization problems. It is composed of two phases: *construction* and *local search*. In the *construction* phase a solution is iteratively constructed through controlled randomization. Then, in the *local search* phase, a local search is applied to the initial solution in order to find a locally optimal solution, and the best overall solution is kept as the result. These two phases are repeated until a stopping criterion is reached. This same procedure is performed in the proposed GNE method, but with different strategies in the *construction* and *local search* phases, as explained in Sect. 5.5.

5.3 Preliminaries

The *Result Diversification Problem* can be stated as a *tradeoff* between finding relevant (similar[1]) elements to the query and diverse elements in the result set. Let $S = \{s_1, \ldots, s_n\}$ be a set of n elements, q a query element and k, $k \leq n$, an integer. Let also the relevance (similarity) of each element $s_i \in S$ be specified by the function $\delta_{sim}(q, s_i)$, $\delta_{sim} : q \times S \to \mathbb{R}^+$, where δ_{sim} is a metric function (i.e., a higher value implies that the element s_i is more relevant to query q). Likewise, let the diversity between two elements $s_i, s_j \in S$ be specified by the function $\delta_{div}(s_i, s_j) : S \times S \to \mathbb{R}^+$. Informally, the problem of result diversification can be described as follows: given a set S and a query q, we want to find $R \subseteq S$ of size $|R| = k$ where each element in R is relevant to q with respect to δ_{sim} and, at the same time, diverse among other elements in R with respect to δ_{div}.

In our model, we represent elements in S using the vector space model. For example, for δ_{sim} and δ_{div} functions the weighted similarity function tf/idf cosine

---

[1] These two terms are used interchangeably throughout the text.

similarity or the Euclidean distance L_2 can be employed to measure the similarity or diversity.

The search for R can be modeled as an optimization problem where there is a *tradeoff* between finding relevant elements to q and finding a diverse result set R. This optimization problem can be divided into two components, one related to similarity, as defined below, and one related to diversity.

Definition 5.1 The **k-similar set** R contains k elements of S that:

$$R = \underset{S' \subseteq S, k=|S'|}{\operatorname{argmax}} sim(q, S')$$

$$\text{where } sim(q, S') = \sum_{i=1}^{k} \delta_{sim}(q, s_i), s_i \in S'$$

in other words, Definition 5.1 finds a subset $R \subseteq S$ of size k with the largest sum of similarity distances among all possible sets of size k in S. Intuitively, $sim(q, S')$ measures the amount of "attractive forces" between q and k elements in S'. Basically, any algorithm that can rank elements in S with respect to δ_{sim} and then extract the top-k elements in the ranked list can evaluate the *k-similar set* problem.

The diversity component is defined as follows:

Definition 5.2 The **k-diverse set** R contains k elements of S that:

$$R = \underset{S' \subseteq S, k=|S'|}{\operatorname{argmax}} div(S')$$

$$\text{where } div(S') = \sum_{i=1}^{k-1} \sum_{j=i+1}^{k} \delta_{div}(s_i, s_j), s_i, s_j \in S'$$

hence, the problem in the above definition selects a subset $R \subseteq S$ of size k that maximizes the sum of inter-element distances amongst elements of R chosen from S. Intuitively, $div(S')$ measures the amount of "repulsive forces" among k elements in S'. The above definition is equivalent to the *Max-Sum Dispersion Problem* [17, 18] (i.e., maximize the sum of distances between pairs of facilities) encountered in operations research literature. For the special case where the distance function δ_{div} is metric, efficient approximate algorithms exist to compute the **k-diverse set** [14].

In both Definitions 5.1 and 5.2, the objective function is defined as a *max-sum problem*, but other measures could be used as well, e.g., *max-min, max-avg, min-max*. The proper measure is very much a matter of the application in hand. Here we develop algorithms specifically for the *max-sum problem* since it seems it has been the most widely accepted among previous approaches (e.g. [2, 4, 7, 12, 16, 37]).

Next, we give the formal definition of our problem: computing a set $R \subseteq S$ of size k with a *tradeoff* between finding k elements that are similar to the query q, defined by Definition 5.1, and finding k elements that are diverse to each other, defined by Definition 5.2. Formally, the problem is defined as follows:

Definition 5.3 Given a *tradeoff* λ, $0 \leq \lambda \leq 1$, between similarity and diversity, the **k-similar diversification set** R contains k elements in S that:

$$R = \underset{S' \subseteq S, k = |S'|}{\mathrm{argmax}} \ \mathcal{F}(q, S')$$

$$\text{where } \mathcal{F}(q, S') = (k - 1)(1 - \lambda) \cdot sim(q, S') + 2\lambda \cdot div(S')$$

Note that, since both components sim and div have different number of elements k and $\frac{k(k-1)}{2}$, respectively, in the above definition the two components are scaled up. The variable λ is a *tradeoff* specified by the user, and it gives the flexibility to return more relevant, or diverse, results for a given query q. Intuitively, $\mathcal{F}(q, S')$ measures the amount of "attractive forces", between q and k elements in S', and "repulsive forces", among elements in S'. Using the same analogy, the above definition selects the "most stable system" R with k elements of S.

In Definition 5.3 there are two special cases for the values of λ. The *k-similar diversification set* problem is reduced to *k-similar set* when $\lambda = 0$, and the result set R depends only on the query q. The second case is when $\lambda = 1$, and the *k-similar diversification set* problem is reduced to finding the *k-diverse set*. In this case, the query q does not play any role to the result set R. Note that our formula is slightly different than the *Max-Sum Diversification Problem* of [12], since it allows us to access both extremes in the search space (diversity only, when $\lambda = 1$, or relevance only, when $\lambda = 0$).

Only when $\lambda = 0$ the result R is straightforward to compute. For any other case, $\lambda > 0$, the associated decision problem is NP-hard, and it can be easily seen by a reduction from the maximum clique problem known to be NP-complete [11]. A brute force algorithm to evaluate the *k-similar diversification set* when $\lambda > 0$ is presented in Algorithm 1. This algorithm tests for every possible subset $R \subseteq S$ of size k to find the highest $\mathcal{F}$ value. Algorithm 1 takes time $O(|S|^k k^2)$, where there are $O(|S|^k)$ possible results S of size k to check, each of which has $O(k^2)$ distances that need to be computed.

Algorithm 1 *k-similar diversification query*

Input: candidate set S and result set size k
Output: result set $R \subseteq S$, $|R| = k$, with the **highest** possible $\mathcal{F}$ value
1: let R be an arbitrary set of size k
2: **for** each set $R' \subseteq S$ of k size **do**
3:     **if** $\mathcal{F}(q, R') > \mathcal{F}(q, R)$ **then**
4:         $R \leftarrow R'$
5:     **end if**
6: **end for**

5.4 Description of Known Methods

In this section we summarize the most relevant known methods for diversifying query results in the literature, for which no extra information other than the relevance and diversity values are used. All methods are casted in terms of the DivDB framework presented in the previous section.

5.4.1 The Swap Method

Perhaps the simplest method to construct the result set R is the *Swap* method [37], which is composed of two phases, as described in Algorithm 2. In the first phase, an initial result R is constructed using the k most relevant elements in S. Then, in the second phase, each remaining element in S is tested, in decreasing order of δ_{sim}, to replace an element from the current solution R. If there is an operation that improves $\mathcal{F}$, then the replace operation that improves $\mathcal{F}$ the most is applied permanently in R. This process continues until every element in the candidate set S is checked.

The $\mathcal{F}$ value of the final result R computed by the *Swap* method is not guaranteed to be optimal, since elements in the candidate set S are analyzed with respect to their δ_{sim} order. That is, this method does not consider the order of δ_{div} values in S, which can result in solutions that do not maximize $\mathcal{F}$.

Algorithm 2 *Swap*

Input: candidate set S and result set size k
Output: result set $R \subseteq S$, $|R| = k$
1: $R \leftarrow \emptyset$
2: **while** $|R| < k$ **do**
3:      $s_s \leftarrow \text{argmax}_{s_i \in S}(\delta_{sim}(q, s_i))$
4:      $S \leftarrow S \setminus s_s$
5:      $R \leftarrow R \cup s_s$
6: **end while**
7: **while** $|S| > 0$ **do**
8:      $s_s \leftarrow \text{argmax}_{s_i \in S}(\delta_{sim}(q, s_i))$
9:      $S \leftarrow S \setminus s_s$
10:     $R' \leftarrow R$
11:     **for** each $s_j \in R$ **do**
12:         **if** $\mathcal{F}(q, \{R \setminus s_j\} \cup s_s) > \mathcal{F}(q, R')$ **then**
13:             $R' \leftarrow \{R \setminus s_j\} \cup s_s$
14:         **end if**
15:     **end for**
16:     **if** $\mathcal{F}(q, R') > \mathcal{F}(q, R)$ **then**
17:         $R \leftarrow R'$
18:     **end if**
19: **end while**

5.4.2 The BSwap Method

The *BSwap* method [37] uses the same basic idea of the *Swap* method for exchanging elements between the candidate set S and the current result R. However, instead of improving $\mathcal{F}$, *BSwap* checks for an improvement in the diversity value div of the current result R, without a sudden drop in δ_{sim} between the exchanged elements.

Algorithm 3 shows the pseudo code for the *BSwap* method. In each iteration, the element in S with the highest δ_{sim} value is swapped with one in R which contributes the least to δ_{div}. If this operation improves div, but without dropping δ_{sim} by a predefined threshold θ, then the result set R is updated (i.e. the two elements are exchanged). This process continues until every element in the candidate set S is tested, or the highest δ_{sim} value in S is below the threshold θ. This later condition is enforced to avoid a sudden drop in δ_{sim} in R. In summary, this method exchanges elements by increasing diversity in expense of relevance.

While very simple, a drawback of *BSwap* is the use of θ. Setting the threshold θ with an appropriate value is a difficult task, and can vary for different datasets and/or queries. If not set properly, the *BSwap* may return results that have less than k elements (if θ is set to a very small value), or with a very low quality in terms of $\mathcal{F}$ (if θ is set to a very large value). As shown in the experimental section, this method also suffers in terms of precision of the results.

Algorithm 3 *BSwap Algorithm*

Input: candidate set S, result set size k, and distance threshold θ
Output: result set $R \subseteq S$, $|R| = k$
1: $R \leftarrow \emptyset$
2: **while** $|R| < k$ **do**
3:      $s_s \leftarrow \text{argmax}_{s_i \in S}(\delta_{sim}(q, s_i))$
4:      $S \leftarrow S \setminus s_s$
5:      $R \leftarrow R \cup s_s$
6: **end while**
7: $s_d \leftarrow \text{argmin}_{s_i \in R}(div(R \setminus s_i))$
8: $s_s \leftarrow \text{argmax}_{s_i \in S}(\delta_{sim}(q, s_i))$
9: $S \leftarrow S \setminus s_s$
10: **while** $\delta_{sim}(q, s_d) - \delta_{sim}(q, s_s) \leq \theta$ **and** $|S| > 0$ **do**
11:     **if** $div(\{R \setminus s_d\} \cup s_s) > div(R)$ **then**
12:         $R \leftarrow \{R \setminus s_d\} \cup s_s$
13:         $s_d \leftarrow \text{argmin}_{s_i \in R}(div(R \setminus s_i))$
14:     **end if**
15:     $s_s \leftarrow \text{argmax}_{s_i \in S}(\delta_{sim}(q, s_i))$
16:     $S \leftarrow S \setminus s_s$
17: **end while**

5.4.3 *The Maximal Marginal Relevance Method*

The Maximal Marginal Relevance (MMR) [2] iteratively constructs the result set R by selecting one new element in S that maximizes the following function:

$$mmr(s_i) = (1 - \lambda)\delta_{sim}(s_i, q) + \frac{\lambda}{|R|} \sum_{s_j \in R} \delta_{div}(s_i, s_j) \qquad (5.1)$$

The MMR method, as illustrated in Algorithm 4, has two important properties that highly influence the chosen elements in the result set R. First, since R is empty in the initial iteration, the element with the highest δ_{sim} value in S is always chosen in R, regardless of its λ value. Second, the remainder $k - 1$ elements are chosen from S that maximize the *mmr* function. Since the result is incrementally constructed by inserting a new element to previous results, the first chosen element has a large influence in the quality of the final result set R. Clearly, if the first element is not chosen properly, then the final result set may have low quality in terms of $\mathcal{F}$. We show in the experimental section that the quality of the results for the MMR method decreases very fast when increasing the λ parameter.

Algorithm 4 MMR

Input: candidate set S and result set size k
Output: result set $R \subseteq S, |R| = k$
1: $R \leftarrow \emptyset$
2: $s_s \leftarrow \text{argmax}_{s_i \in S}(mmr(s_i))$
3: $S \leftarrow S \setminus s_s$
4: $R \leftarrow s_s$
5: **while** $|R| < k$ **do**
6:      $s_s \leftarrow \text{argmax}_{s_i \in S}(mmr(s_i))$
7:      $S \leftarrow S \setminus s_s$
8:      $R \leftarrow R \cup s_s$
9: **end while**

5.4.4 *The Motley Method*

Similarly to MMR, the *Motley* approach [16] also iteratively constructs the result set R (as illustrated in Algorithm 5).[2] In the first iteration, the element with the highest δ_{sim} value is inserted in R. Then, an element from S is chosen to be included in R if it has δ_{div} value greater than a predefined threshold θ' to every element already

---

[2] This is a simplified version of the Buffered Greedy Approach [16]. In our preliminary experiments, the results of both approaches were very similar, except that the one described here is several orders of magnitude faster than the Buffered Greedy.

inserted in R. If such condition is not satisfied, then the element is discarded from S and the next element with the highest δ_{sim} value in S is evaluated. This process repeats until the result set R has k elements, or S has no more elements.

Like the MMR method, the *Motley* approach is also affected by the initial choice of element. Similarly to the *BSwap*, *Motley* uses a threshold parameter to check the amount of diversity in the result set.

Algorithm 5 *Motley*

Input: candidate set S and result set size k
Output: result set $R \subseteq S, |R| = k$
1: $s_s \leftarrow \text{argmax}_{s_i \in S}(\delta_{sim}(q, s_i))$
2: $S \leftarrow S \setminus s_s$
3: $R \leftarrow s_s$
4: **while** $|R| < k$ **do**
5:     $s_s \leftarrow \text{argmax}_{s_i \in S}(\delta_{sim}(q, s_i))$
6:     $S \leftarrow S \setminus s_s$
7:     **if** $\delta_{div}(s_r, s_s) \geq \theta', \forall s_r \in R$ **then**
8:         $R \leftarrow R \cup s_s$
9:     **end if**
10: **end while**

5.4.5 The Max-Sum Dispersion Method

The Max-Sum Dispersion (MSD) method [12] is based on the 2-approximation algorithm proposed in [14] for the *Max-Sum Dispersion Problem*. The MSD method employs a greedy approach to incrementally construct the result set R by selecting a pair of elements that are relevant to the query and diverse to each other. More formally, in each iteration of the MSD method, the pair of elements $s_i, s_j \in S$ that maximize the following function is chosen to be part of the result set R:

$$msd(s_i, s_j) = (1 - \lambda)(\delta_{sim}(q, s_i) + \delta_{sim}(q, s_j)) + 2\lambda\delta_{div}(s_i, s_j) \qquad (5.2)$$

Since a pair of elements is selected in each iteration of the MSD method, the above description applies only when the value for k is even. If k is odd, in the final phase of the MSD method an arbitrary element in S is chosen to be included in the result set R. The pseudo code for the MSD method is shown in Algorithm 6.

Algorithm 6 MSD

Input: candidate set S and result set size k
Output: result set $R \subseteq S, |R| = k$
1: $R \leftarrow \emptyset$
2: **while** $|R| < \lfloor k/2 \rfloor$ **do**
3:    $\{s_i, s_j\} \leftarrow \text{argmax}_{s_i, s_j \in S}(msd(s_i, s_j))$
4:    $R \leftarrow R \cup \{s_i, s_j\}$
5:    $S \leftarrow S \setminus \{s_i, s_j\}$
6: **end while**
7: **if** k is odd **then**
8:    choose an arbitrary object $s_i \in S$
9:    $R \leftarrow R \cup s_i$
10: **end if**

5.4.6 The Clustering-Based Method

The Clustering-Based Method (CLT) [21] is based on clustering techniques, and its pseudo code is shown in Algorithm 7. In its first phase the k-medoid algorithm with δ_{div} as distance function is employed in the elements in S to generate k clusters $C = \{c_1, c_2, \ldots, c_k\}$. Then, one element from each cluster in C is included in the result set R. Several strategies on selecting an element from each cluster can be employed. For example, one can choose the element with the highest δ_{sim} value, or also the medoid element of each cluster C.

Since it is not possible to incorporate the *tradeoff* λ in the clustering step of the CLT method, the result set R only depends on the strategy employed for selecting an element in each cluster. Thus, for low values of λ, the result set R will be of better quality if the element with the highest δ_{sim} is selected. For higher λ values selecting the medoid element of each cluster might be more suitable since the medoids typically have higher δ_{div} values to each other. In our experiments we only consider the medoid-based strategy since it typically gives more diverse results.

Algorithm 7 CLT

Input: candidate set S and result set size k
Output: result set $R \subseteq S, |R| = k$
1: $R \leftarrow \emptyset$
2: let $C = \{c_1, c_2, ..., c_k\}$ be the result of the k-medoid algorithm, using δ_{div}
3: **for** each cluster $c_i \in C$ **do**
4:    let s_i be a selected element in c_i
5:    $R \leftarrow R \cup s_i$
6: **end for**

5.5 New Proposed Methods

We proceed with the presentation of two new approaches for diversifying query results, the *Greedy Marginal Contribution* (GMC) and the *Greedy Randomized with Neighborhood Expansion* (GNE) methods. Both methods employ the same function to rank elements regarding their marginal contribution to the solution. An important difference between the two methods is that in the GMC the element with the highest partial contribution is always chosen to be part of the solution, while in GNE an element is randomly chosen, among the top ranked ones, to be included in the solution.

5.5.1 The GMC Method

The GMC method incrementally builds the result R by selecting the element with the highest maximum marginal contribution (mmc) to the solution constructed so far. In each iteration, the GMC method ranks the elements in the candidate set S using the following function:

$$mmc(s_i) = (1-\lambda)\delta_{sim}(s_i, q) + \frac{\lambda}{k-1} \sum_{s_j \in R_{p-1}} \delta_{div}(s_i, s_j) + \frac{\lambda}{k-1} \sum_{\substack{l=1 \\ s_j \in S - s_i}}^{\substack{l \leq k-p}} \delta^l_{div}(s_i, s_j)$$

(5.3)

where R_{p-1} is the partial result of size $p-1$, $1 \leq p \leq k$, and $\delta^l_{div}(s_i, s_j)$ gives the lth largest δ_{div} value in $\{\delta^l_{div}(s_i, s_j) : s_j \in S - R_{p-1} - s_i\}$. Using the above function we can compute the maximum contribution value of s_i to $\mathcal{F}$ using the elements already inserted in the partial result R_{p-1} (first two components) and the remaining $k - p$ elements in S that could be inserted in R (third component). For the third component, the diversity value of s_i is upper bounded using the highest $k - p$ values for δ_{div} that is formed using the p elements in S. Thus, the mmc function is employed to rank every element in S using the constructed result R_{p-1} so far, and the remaining elements that can be included in the final result R.

Algorithm 8 GMC

Input: candidate set S and result set size k
Output: result set $R \subseteq S$, $|R| = k$
1: $R_0 \leftarrow \emptyset$
2: **for** $p \leftarrow 1$ to $p = k$ **do**
3:     $s_i \leftarrow \text{argmax}_{s_i \in S}(mmc(s_i))$
4:     $R_p \leftarrow R_{p-1} \cup s_i$
5:     $S \leftarrow S \setminus s_i$
6: **end for**
7: $R \leftarrow R_p$

The intuition behind mmc is the following: When the current result set R_0 is empty, then the elements in S are ranked based on their δ_{sim} values and their relationships, defined by δ_{div}^l, with other elements in S. In this way, the mmc function gives higher preference to elements that are very relevant to q and, at the same time, are very diverse to other elements in the candidate set S, regardless of R_0; whenever elements are inserted in R_p, then mmc also considers the diversity w.r.t. elements in R_{p-1}.

Compared to MMR (Sect. 5.4.3), the GMC method uses a different function to rank the elements in the candidate set S. Since R is empty in the first step of the MMR, the diversity component does not play any role in selecting the first element in R. On the other hand, in GMC the selection of the first result element is based on the maximum marginal contribution value that the element can make to $\mathcal{F}$. Note that mmc combines the relevance of a new element, its diversity to the already inserted elements in the result, and its maximum contribution to $\mathcal{F}$ considering the remaining elements in S.

Algorithm 8 shows the pseudo code for the GMC method. In the first step of the method the sum of δ_{div} in mmc is zero, since the result set R_0 is empty. Thus, only the δ_{sim} and δ_{div}^l components are considered. In the following iterations mmc considers the elements in R_{p-1}, and the value for the δ_{div}^l component is updated. As more elements are inserted in the result set R, the values of the second and third components in the mmc function increases and decreases, respectively. In the last iteration ($p = k$), the third component δ_{div}^l is null, since there is only one element to be considered in R.

The δ_{div}^l values for each element $s_i \in S$ are computed only once, and it is done in the first iteration of the GMC method. For each entry $s_i \in S$, a set of pairs $< s_j, \delta_{div}(s_i, s_j) >$ is maintained, where s_j is the element that belongs to one of the δ_{div}^l and its corresponding value $\delta_{div}(s_i, s_j)$. This set begins with $k - 1$ entries and in each iteration decreases by 1. The set of pairs for each element in S is updated in each iteration of the method. Two cases can occur:

1. the first is when there is an element $s_i \in S$ that has the pair $< s_i', \delta_{div}(s_i, s_i') >$ in its set of pairs, where element s_i' is the element previously inserted in the result set R_p. Then the pair $< s_i', \delta_{div}(s_i, s_i') >$ is dropped from the set of pairs from s_i;
2. the second case happens when there is an element $s_i \in S$ that does not have the pair $< s_i', \delta_{div}(s_i, s_i') >$ in its set of pairs. In this case, the pair with the lowest $\delta_{div}(s_i, s_j)$ is removed from the set.

5.5.2 The GNE Method

Our second proposed method uses the GRASP (Greedy Randomized Adaptive Search Procedure) [10] technique for diversifying query results. To the best of our knowledge, this is the first randomized solution proposed for the diversification problem. Different from the GMC that always selects the top element in the rank, in each iteration of the GNE algorithm a random element is chosen among the top ranked ones. Algorithm 9 illustrates the general GRASP algorithm with i_{max} iterations. The GNE algorithm has two phases, the *Construction Phase* and the *Local Search Phase*, which are described below.

Algorithm 9 *GRASP*

Input: candidate set S and result set size k
Output: result set $R \subseteq S$, $|R| = k$
1: $R \leftarrow \emptyset$
2: **for** $i \leftarrow 0$ to $i < i_{max}$, $i \leftarrow i + 1$ **do**
3:     $R' \leftarrow$ GNE-Construction()
4:     $R' \leftarrow$ GNE-LocalSearch(R')
5:     **if** $\mathcal{F}(q, R') > \mathcal{F}(q, R)$ **then**
6:         $R \leftarrow R'$
7:     **end if**
8: **end for**

5.5.2.1 GNE Construction Phase

In the construction phase, a greedy randomized ranking function chooses an element to be inserted in R. This ranking function ranks the elements in S according to the mmc function, as described in Eq. 5.3. In each of the K iterations of this phase, a list, called *Restricted Candidate List* (RCL), is constructed with elements with the highest individual contributions with respect to mmc. In this phase, the GNE method incrementally builds the result set R by randomly selecting an element (which may not be the element with the highest contribution) in the RCL. Therefore, in each iteration all elements in S are re-ordered based on mmc, and only the top elements are stored in the RCL. This iterative process continues until a solution with k elements is constructed.

Algorithm 10 *GNE-Construction*

Output: a candidate solution $R \subseteq S$, $|R| = k$
1: $R_0 \leftarrow \emptyset$
2: **for** $p \leftarrow 1$ to $p = k$ **do**
3:     $s_{max} \leftarrow argmax_{s_i \in S}(mmc(s_i))$
4:     $s_{min} \leftarrow argmin_{s_i \in S}(mmc(s_i))$
5:     $RCL \leftarrow \{s_i \in S | mmc(s_i) \geq s_{max} - \alpha(s_{max} - s_{min})\}$
6:     $s_i \leftarrow random(RCL)$
7:     $R_p \leftarrow R_{p-1} \cup s_i$
8:     $S \leftarrow S \setminus s_i$
9: **end for**
10: $R \leftarrow R_p$

The selection of the elements with the highest contributions to be included in the RCL can be based on a fixed number of candidates, or on a predefined threshold.

Table 5.1 Description of the methods evaluated

Abbreviation	Method name	Construction of R
Swap	*Swap*	Exchanging
BSwap	*BSwap*	Exchanging
MMR	*Maximal Marginal Relevance*	Incremental
Motley	*Motley*	Incremental
MSD	*Max-Sum Dispersion*	Incremental
CLT	*Clustering*	Exchanging
GMC	*Greedy Marginal Contribution*	Incremental
GNE	*GRASP with Neighbor Expansion*	Random, meta-heuristic
Random	*Random*	Random

The intuition behind limiting the size of RCL is that not every element in S contributes equally to the result set. Therefore, the method performs randomized search over the elements in the RCL that can more contribute to the solution. In our implementation, we use the first approach to restrict the number of candidates in the RCL, as shown in Algorithm 10.

Parameter α controls how greedy and random the *Construction* phase is. For $\alpha = 0$, the *Construction* corresponds to a pure greedy construction procedure, working exactly the same as the GMC method. On the other hand, for $\alpha = 1$ the construction phase produces a random construction, regardless of the mmc ranking values.

Algorithm 11 *GNE-LocalSearch*

Input: candidate solution R of size k
Output: a candidate solution $R \subseteq S$, $|R| = k$
1: $s_{max} \leftarrow argmax_{s_i \in S}(mmc(s_i))$
2: $s_{min} \leftarrow argmin_{s_i \in S}(mmc(s_i))$
3: $RCL \leftarrow \{s_i \in S | mmc(s_i) \geq s_{max} - \alpha(s_{max} - s_{min})\}$
4: **for** each $s_i \in R$ **do**
5:     $R' \leftarrow R$
6:     **for** each $s_j \in R, s_j \neq s_i$ **do**
7:         **for** $l \leftarrow 1$ to $l = k - 1$ **do**
8:             $s_i^l \leftarrow \{s_i' \in S | \delta_{div}^l(s_i, s_i')\}$
9:             **if** $s_i^l \notin R$ **then**
10:                 $R'' \leftarrow R' - s_j + s_i^l$
11:                 **if** $\mathcal{F}(q, R'') > \mathcal{F}(q, R')$ **then**
12:                     $R' \leftarrow R''$
13:                 **end if**
14:             **end if**
15:         **end for**
16:     **end for**
17:     **if** $\mathcal{F}(q, R') > \mathcal{F}(q, R)$ **then**
18:         $R \leftarrow R'$
19:     **end if**
20: **end for**

Since both GNE and GMC methods use the same greedy strategies to measure the individual contribution of an element to the result set, performing random selections during the construction phase provides the contribution of trying different elements in each multistart construction.

5.5.2.2 GNE Local Search Phase

Starting with the current solution R of size k provided by the construction phase, the local search progressively improves it by applying a series of local modifications in the neighborhood of R. Repeated runs of the construction phase yield diverse starting solutions for the local search.

The proposed *local search* performs *swaps* between elements in the result set R and their most diverse element. If this procedure improves the current solution R, then a locally optimal solution is found. This is accomplished by a *Neighborhood Expansion* that explores the most diverse elements of each entry in the result set. This expansion phase is performed as follows: for each element in the result set R, the $k - 1$ most diverse elements are computed; then, for each of the remaining elements in R, a swap operation is performed with every element among the $k - 1$ most diverse elements. If in any of these swap operations a local solution is found, then this partial solution is made the best optimal solution. The pseudo code for the *local search* is shown in Algorithm 11.

5.5.3 Complexity Analysis

The time complexity of GMC is $O(kn^2)$, since we have to compute the *mmc* for each element in S, which is $O(kn)$. GNE has two parts: Algorithm 10 which has time complexity $O(kn^2)$, and Algorithm 11 which is $O(k^3)$ (δ^l_{div} is computed only once). Nevertheless, the GNE method runs i_{max} times, which makes it slower in practice than the GMC method.

5.6 Experimental Evaluation

We proceed with the experimental evaluation of all the methods previously described in this chapter. The method's abbreviation, name and strategy employed to construct the result set R appear in Table 5.1. As a baseline for our comparisons, we also included the results for a random method, called *Rand*, which simply chooses the result set R with the highest $\mathcal{F}$ among 1,000 random runs. The *Rand* method sets a base line for comparisons where a specific method needs to outperform random guessing.

Table 5.2 Datasets statistics

Dataset	# of elements	Maximum # of attributes	Average # of attributes	δ_{sim}	δ_{div}
trajectories	17,621	234,744	2,701	$1 - L_2(20, 40..200)$	$L_2(13, 26..260)$
docs	25,276	15,116	363	$1 - cosine(terms)$	$cosine(docs)$
dblp	2,991,212	68	7	$1 - cosine(terms)$	$cosine(title)$

5.6.1 Setup

Three datasets from several different domains were used to evaluate each method with respect to running time, precision and $\mathcal{F}$ of the results. The datasets used in our experiments are: *trajectories*, *docs* and *dblp*. *trajectories* [40] contains 23,793,860 GPS locations generated from 178 users in a period of over three years. In our experiments, we use 17,621 trajectories which represent the continuous movements of the 178 users. Besides performing experiments using a trajectory dataset, which is the main focus of this book, we also employed five other datasets from various domains. The main idea of using other datasets is to show that the DivDB framework is generic and can also be applied in any domain that the metric functions δ_{sim} and δ_{div} can be defined. Dataset *docs* has 25,960 news articles obtained from the TREC-3 collection [35], while *dblp* is a set of 2,991,212 publication entries extracted from the DBLP Bibliography Server [22]. In order to generate duplicate entries in a controlled way, and thus test the diversity in the result while increasing the λ parameter, each entry in the *dblp* dataset contains the author's name and the publication title. A DBLP publication with multiple authors thus has multiple entries in *dblp*, one entry per author, sharing the same publication title. The detailed statistics of all datasets are summarized in Table 5.2.

In the experiments we employed the same distance function for both δ_{sim} and δ_{div}. The Euclidean distance for *trajectories*, and the *cosine* similarity distance for both *docs* and *dblp*. For the *trajectories* dataset δ_{sim} was computed using part (one quarter) of the feature vector, while for δ_{div} the whole feature vector was used. For instance, as for the *trajectories* dataset, δ_{sim} and δ_{div} use 10 and 20, respectively, sparse locations of the trajectory query, as shown in Table 5.2. For the *docs* dataset 10 random terms were used for δ_{sim} while the whole document was utilized for δ_{div}. The same process was employed in *dblp*, but using a few terms extracted from the publication titles.

To generate a query set we randomly selected 100 queries from each dataset. The extra parameters θ and θ' for *BSwap* and *Motley*, respectively, were set to 0.1, which achieved, on average, the best performance in the experiments. For the GNE we set $i_{max} = 10$ and $\alpha = 0.01$. Clearly, GNE returns better results when increasing the values for those two parameters, but at a cost of higher running times. For example, when increasing the α parameter, the GNE method slowly converges to good solutions since there are many more elements in the candidate set S to be considered.

Table 5.3 Parameters tested in the experiments

Parameter	Range (*default*)
Tradeoff λ values	0.1, 0.3, 0.5, **0.7**, 0.9
Candidate set size $n = |S|$	200, **1,000**, 2,000, 3,000, 4,000
Result set size $k = |R|$	**5**, 15, 25, 35
# of sample queries	**100**

From our preliminary experiments the above values for the GNE parameters provided a good tradeoff between *quality* and running time.

Table 5.3 summarizes the testing parameters and their corresponding ranges, in which the default value is marked in bold font. To obtain the candidate set S we employed an *off-the-shelf* retrieval system. That is, for a particular query sample, S contains the top-n elements in the dataset using the ranking function δ_{sim}. Generally speaking, each group of experimental studies is partitioned into two parts: qualitative analysis in terms of $\mathcal{F}$ and precision of each method with respect to the *optimal result set*; and scalability analysis of $\mathcal{F}$ and running time when increasing the query parameters.

5.6.2 Qualitative Analysis

To measure a method's precision we need to compare it against the optimal result set (the one that maximizes $\mathcal{F}$). However, due to the problem complexity, we can only compute the optimal result set for a limited range of the query parameters. Using the brute force algorithm (Algorithm 1), we computed the optimal result set for each dataset with parameters $S = 200$, $k = 5$ and λ varying from 0.1 to 0.9. The precision is measured by counting the number of common elements in the result set from Algorithm 1 and the result set for each method using the same query parameters; clearly, the higher the precision, the better the method performs.

Figure 5.1 shows the average precision of each method for different values of λ (preference to diversity increase as λ increases). Note that, the precision of *all* methods, except GNE and GMC, typically decreases when increasing λ. Interestingly, four of the previous approaches (*Swap*, *BSwap*, *Motley* and CLT) performed worse than the *Rand* method when λ is greater than 0.5. That is, if diversity is more prominent than relevance, these methods are easily outperformed.

GNE and GMC outperformed all the methods and showed almost constant precision regarding λ. Only in the case where $\lambda = 0.1$ (i.e., when relevance is much more important than diversity), the MMR method have precision similar to GNE/GMC for three datasets (*faces*, *nasa* and *colors*). Previous methods are influenced by the first most relevant element in their result set, which affects the result precision when diversity has a higher weight than relevance. In contrast, the variation of λ does not

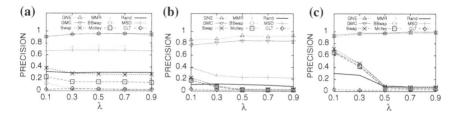

Fig. 5.1 Average precision versus *tradeoff* λ values (**a**) *trajectories* (**b**) *docs* (**c**) *dblp*

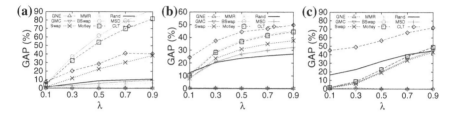

Fig. 5.2 Average gap versus *tradeoff* λ values (**a**) *trajectories* (**b**) *docs* (**c**) *dblp*

affect the precision of GNE and GMC. The average precision for GNE and GMC
was at least 75 %.

The precision of GNE is better than GMC in all datasets tested, except for the
dblp dataset. This is because GNE selects elements from RCL using a randomized
approach, that is, GNE may not always select the element with the highest contribution to the result set. Furthermore, given an initial solution the GNE also uses the
local search phase that aims at finding better solutions.

We also compared the $\mathcal{F}$ values of each method against the $\mathcal{F}$ value of the *optimal
result set*. This experiment is intended to measure how close the $\mathcal{F}$ for a result set
is to the maximum possible $\mathcal{F}$ value. Figure 5.2 depicts the "gap" s in $\mathcal{F}$, i.e., the
difference between the $\mathcal{F}$ of a particular method and the optimal $\mathcal{F}$, divided by the
optimal. The *gap* for GNE and GMC is close to zero for every dataset and any value
of λ, which indicates that these two methods constructed results sets with (very close)
maximum $\mathcal{F}$ values. As for the other methods the *gap* increases with λ. Again, for
for high values of λ the *BSwap*, *Swap*, *Motley* and CLT are typically outperformed
by the *Rand* method.

The effectiveness of GMC and GNE is also depicted in Table 5.4 which illustrates
the result sets for four example queries, using the *dblp* dataset ($k = 5$ and $S = 200$)
with $λ = 0$ (*no diversification*), $λ = 0.3$ (*moderate diversification*) and $λ = 0.7$
(*high diversification*). Only GNE is shown since the results for GMC are similar.
When $λ = 0$ the result set contains the top-5 elements of S ranked with the δ_{sim}
scoring function. Since the *dblp* dataset contains several duplicate entries, the result
sets with *no diversification* contain several duplicate elements (we also show the
author names for each entry as an *identification* for that entry). As λ increases, fewer
duplicates are found in the result set, while the elements in the result set "cover"

Table 5.4 Result sets for *dblp* dataset: top-5 results ($\lambda = 0$), and two diversified results ($\lambda = 0.3$ and 0.7) produced by GNE

	Top-k: no diversification ($\lambda = 0$)	GNE: moderate diversification ($\lambda = 0.3$)	GNE: high diversification ($\lambda = 0.7$)
q_3 = *data mining*			
1	D. Olson: **data mining**	K. Rihaczek: **data mining**	S. Olafsson: operations research and **data mining**
2	J. Kunz: **data mining**	J. Kunz: **data mining**	S. Kaya: privacy in spatiotemporal **data mining**
3	K. Rihaczek: **data mining**	D. Olson: **data mining**	M. Steinbach: top 10 algorithms in **data mining**
4	S. Morishita: **data mining** tools for the...	B. Toursel: distributed **data mining**	P. Perner: **data mining** concepts and techniques
5	Y. Ohya: **data mining** tools for the...	R. Agrawal: whither **data mining**?	C. G-Carrier: personalizing e-commerce with **data mining**
q_1 = *database systems*			
1	Y. Theodoridis: trajectory **database systems**	C. Date: an introduction to **database systems**	R. Deshpande: an electronic **database** delivery **system**
2	I. Ntoutsi: trajectory **database systems**	J. Rothnie: distributed **database systems**	T. Xu: pbmice: an integrated **database system** of piggyBac...
3	N. Pelekis: trajectory **database systems**	N. Pelekis: trajectory **database systems**	C. Matheus: **systems** for knowledge discovery in **databases**
4	E. Frentzos: trajectory **database systems**	M. Wu: the dbo **database system**	D. Dimitroff: an intelligent image **database system**
5	A. Elmagarmid: videotext **database systems**	J. Gray: parallel **database systems** 101	F. Ozcan: metu interoperable **database system**
q_4 = **medical image**			
1	O. Pianykh: **medical image** enhancement	P. Wang: **medical image** processing	M. Moshfeghi: elastic matching multimodality **medical images**
2	J. Tyler: **medical image** enhancement	R. Deklerck: segmentation of **medical images**	S. Zagal: segmentation of **medical images** by region growing
3	M. Trifas: **medical image** enhancement	C. Meinel: compression of **medical images**	C. Chang: detect. and restor. tampered **medical image**

(continued)

Table 5.4 (continued)

	Top-k: no diversification ($\lambda = 0$)	GNE: moderate diversification ($\lambda = 0.3$)	GNE: high diversification ($\lambda = 0.7$)
4	C. Meinel: compression of **medical images**	J. Tyler: **medical image** enhancement	C. Faloutsos: similarity searching in **medical image** databases
5	S. Hludov: compression of **medical images**	C-H. Lin: quality of compressed **medical images**	P. Radeva: discriminant snakes for 3d reconstr. **medical images**
q_2 = **nearest neighbor**			
1	H. Alt: the **nearest neighbor**	H. Alt: the **nearest neighbor**	W. Tiller: algorithm for finding all k **nearest neighbors**
2	G. Yuval: finding **nearest neighbors**	G. Yuval: finding **nearest neighbors**	L. Wu: stepwise **nearest neighbor** discriminant analysis
3	F. Yao: on **nearest-neighbor** graphs	C. Domeniconi: **nearest neighbor** ensemble	B. Pittel: the random bipartite **nearest neighbor** graphs
4	M. Paterson: on **nearest-neighbor** graphs	P. Grother: fast impl. **nearest neighbor** classifier	R. Meiche: a hardware accelerator for k-th **nearest neighbor**...
5	D. Eppstein: on **nearest-neighbor** graphs	M. Paterson: on **nearest-neighbor** graphs	X. Yu: research on an adaptive k-**nearest neighbors** classifier

many more subjects (defined by terms in the publication name). Observe that the result set with *high diversification* contains elements that all have the query terms, as well as other terms indicating that the publication is related to different subjects among the other publications in the result set (i.e., hardware accelerator, operations research, privacy in spatio-temporal).

5.6.3 Scalability

We next examine the scalability of the methods with respect to $\mathcal{F}$ values and running time while varying the query parameters λ (from 0.1 to 0.9), k (from 5 to 35) and S (from 200 to 4,000). We omit the results for the *trajectories* dataset since they are similar to *colors*. Figure 5.3 shows the results for $\mathcal{F}$, with $k = 5$ and $S = 1,000$, when increasing λ. Regardless λ, for GNE and GMC the $\mathcal{F}$ values are the highest among

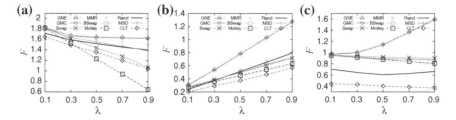

Fig. 5.3 Average $\mathcal{F}$ value versus *tradeoff* λ values (**a**) *colors* (**b**) *docs* (**c**) *dblp*

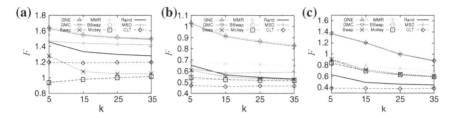

Fig. 5.4 Average $\mathcal{F}$ value versus result set k size (**a**) *colors* (**b**) *docs* (**c**) *dblp*

all other methods. In the *colors* dataset, $\mathcal{F}$ decreases when increasing λ because the δ_{div} values among elements in S are not as high as their δ_{sim} values. The difference between $\mathcal{F}$ values for GNE and GMC and the other methods are greater as the value of λ increases. This is due to the fact that in all previous methods the diversification of the result set is performed giving preference on the ranking order that the elements are stored in S, which are ordered by δ_{sim}. In other words, as opposed to GNE and GMC, all previous approaches do not diversify results for arbitrary values of λ.

Figure 5.4 shows $\mathcal{F}$ when increasing k, with $\lambda = 0.7$ and $S = 1,000$. The $\mathcal{F}$ values decrease when increasing k because the δ_{div} values among elements in R tend to be smaller for the same candidate set S. In addition, more elements are considered in the computation of $\mathcal{F}$, which in turn decreases $\mathcal{F}$. Regardless of the value of k, the $\mathcal{F}$ values for GNE and GMC are again higher among all other methods and datasets.

Figure 5.5 depicts the results when increasing the size of the candidate size S, with $k = 5$ and $\lambda = 0.7$. For the *dblp* dataset the values of $\mathcal{F}$ are constant for all methods, except for the *Rand* and CLT. This is because in the *dblp* dataset there are many more elements that do not have a high value of δ_{sim} and are included in the candidate set S. In this case, a good tradeoff between relevance and diversity is found for R with size $k = 5$ and small size of S (e.g., $S = 200$). For the other two datasets, *colors* and *docs*, a larger size of S leads to better result set R for the required tradeoff ($\lambda = 0.7$).

Interestingly, for the *docs* and *dblp* datasets the values of $\mathcal{F}$ in the CLT and *Rand* methods tend to decrease as the size of S increases. This is because in both methods a fixed number of runs, i.e. 1,000, is performed in order to select the best solution, but as the number of candidates increases more runs are needed for these two methods

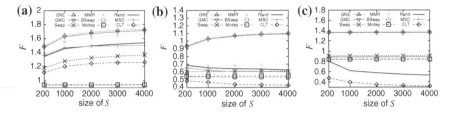

Fig. 5.5 Average $\mathcal{F}$ value versus candidate set S size (**a**) *colors* (**b**) *docs* (**c**) *dblp*

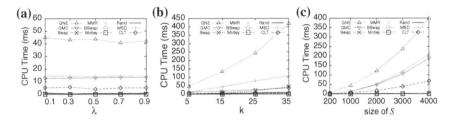

Fig. 5.6 Average running time versus (**a**) λ (**b**) k and (**c**) S for the *dblp* dataset

to converge to good solutions. As a result, the CLT and *Rand* methods do not scale well for large sizes of S. On the other hand, the other methods do not exhibit this behavior since the elements in S are analyzed only once.

Overall, for all datasets and query parameters the highest $\mathcal{F}$ values were achieved by the GNE and GMC methods. GNE provides slightly better $\mathcal{F}$ values (up to 2 % higher than the GMC). Since the GNE approach randomly selects an element in the RCL to be included in the result set R, its performance depends not only on the size of the RCL (defined by α), but also on the size of S. This is because i_{max} is affected by α and the distribution of contribution of each element in S (measured using mmc). Therefore, increasing S for the same α also increases the size of RCL, which enables the GNE method to *slowly* converge to better result sets. One possible approach to cope with this, without interfering with GNE's running time, is to limit the size of RCL and/or to increase the value for the i_{max} parameter.

And finally, Fig. 5.6 depicts the running time (in *ms*) for the *dblp* dataset when varying λ ($k = 5$ and $S = 1,000$), k ($\lambda = 0.7$ and $S = 1,000$) and S ($k = 5$ and $\lambda = 0.7$). Since the same behavior was observed for the other two datasets, here we only discuss the running time for the *dblp*. As expected, the running time is not affected when increasing λ, as shown in Fig. 5.6a. This parameter does not interfere on how a method operates, but only dictates the selection of elements in the result set. The GNE method has the highest running time among all methods, since it executes a few runs over the candidate set S and also has an expensive local search phase.

In Fig. 5.6b, the running time increases proportionally to the size of k, since more iteration is performed over S. The *BSwap* and *Motley* exhibit the slowest increase. The running time for GNE and GMC increases for large k since both compute, for each element in S, the $k - 1$ elements in S that have the highest δ_{div} values.

Nevertheless, GMC is competitive having a similar behavior to *Swap* for larger k values compared to other methods. As for GNE, the larger increase is because the number of runs and exchanges in the local search phase are both proportional to k.

In Fig. 5.6c, the running time of all methods, except for *BSwap* and *Motley*, increases with S since all elements in S have to be checked. Since *BSwap* and *Motley* have early stopping conditions, defined by the threshold conditions θ and θ', respectively, their performance is not directly influenced by the size of S but by their threshold values. Again, GNE had the highest running times since it performs several runs, which depend on k and S. GMC and MSD have again similar behavior when increasing S.

5.6.4 Experimental Summary

The *BSwap*, *Swap*, *Motley* and CLT methods were outperformed by the *Rand* approach in terms of $\mathcal{F}$ values in almost all experiments, which indicates that a simple random selection returns better results than these four methods. As expected, among all methods the CLT is the one that achieved the lowest $\mathcal{F}$ values for the majority of the experiments. This last result corroborates our initial claim that clustering techniques do not apply well for query result diversification. Besides having an extra parameter that is not straightforward to tune, both *BSwap* and *Motley* provided the lowest precision and $\mathcal{F}$ values in their results. Among the previous methods tested (i.e., excluding our proposed GNE and GMC methods), MMR was typically better in most of the experiments.

The GNE and GMC methods consistently achieved better results in terms of precision, *gap* and $\mathcal{F}$ values for every dataset and query parameter tested. Among them, GNE generally provided the best results, since it performs several runs in a controlled way. However, this affects its running time; GNE was the slowest method among all competitors. GMC achieved running times which are similar to MSD and CLT. Overall, GMC provided high quality results with acceptable running times. If the user is willing to further improve quality at the expense of higher running time, then GNE should be preferred over GMC.

5.7 Final Remarks

In this chapter, we describe the DivDB, a simple yet powerful framework for evaluation and optimization of methods for diversifying query results. One advantage of the DivDB framework is that users can set the *tradeoff* between finding the most relevant elements to the query and finding diverse elements in the result set. Another advantage is that we can quantitatively compare different methods implemented in the DivDB framework. We then describe several known methods, as well as two new methods, for diversifying query results. The two new proposed methods, named

GMC and GNE, construct the result set in an incremental way using a function that computes the contribution of each candidate element considering its relevance to the query and its diverse value to the elements in the current result set and to other elements in the candidate set. In GMC the element with the highest contribution value is always included in the result set, while in GNE a randomized procedure is employed to choose, among the top highest elements, the one to be included in the result set. We show a thorough experimental evaluation of the various diversification methods using real datasets. GMC and GNE methods outperformed all known methods for any *tradeoff* values between relevance and diversity in the result set.

References

1. Agrawal, R., Gollapudi, S., Halverson, A., Ieong, S.: Diversifying search results. In: Proceedings of the ACM International Conference on Web Search and Data Mining (WSDM), pp. 5–14. ACM (2009). http://dx.doi.org/10.1145/1498759.1498766
2. Carbonell, J., Goldstein, J.: The use of MMR, diversity-based reranking for reordering documents and producing summaries. In: Proceedings of the ACM SIGIR International Conference on Research and Development in Information Retrieval, pp. 335–336. ACM (1998). http://dx. doi.org/10.1145/290941.291025
3. Carterette, B.: An analysis of NP-completeness in novelty and diversity ranking. Information Retrieval **14**(1), 89–106 (2011). http://dx.doi.org/10.1007/s10791-010-9157-1
4. Chapelle, O., Metlzer, D., Zhang, Y., Grinspan, P.: Expected reciprocal rank for graded relevance. In: Proceedings of the ACM International Conference on Information and Knowledge Management (CIKM), pp. 621–630 (2009). DOI http://dx.doi.org/10.1145/1645953.1646033
5. Chen, H., Karger, D.: Less is more: probabilistic models for retrieving fewer relevant documents. In: Proceedings of the ACM SIGIR International Conference on Research and Development in Information Retrieval, pp. 429–436. ACM (2006). DOI http://dx.doi.org/10.1145/ 1148170.1148245
6. Clarke, C., Kolla, M., Cormack, G., Vechtomova, O., Ashkan, A., Büttcher, S., MacKinnon, I.: Novelty and diversity in information retrieval evaluation. In: Proceedings of the ACM SIGIR International Conference on Research and Development in Information Retrieval, pp. 659–666. ACM (2008). DOI http://dx.doi.org/10.1145/1390334.1390446
7. Coyle, M., Smyth, B.: On the importance of being diverse. In: Z. Shi, Q. He (eds.) Intelligent Information Processing II, IFIP International Federation for Information Processing, vol. 163, pp. 341–350. Springer US (2005). DOI http://dx.doi.org/10.1007/0-387-23152-8_43
8. Demidova, E., Fankhauser, P., Zhou, X., Nejdl, W.: DivQ: Diversification for keyword search over structured databases. In: Proceedings of the ACM SIGIR International Conference on Research and Development in Information Retrieval, pp. 331–338. ACM (2010). DOI http:// dx.doi.org/10.1145/1835449.1835506
9. Drosou, M., Pitoura, E.: Diversity over continuous data. IEEE Data Eng. Bull. **32**(4), 49–56 (2009)
10. Feo, T.A., Resende, M.G.C.: Greedy randomized adaptive search procedures. J. of Global Optimization **6**(2), 109–133 (1995). DOI http://dx.doi.org/10.1007/BF01096763
11. Garey, M.R., Johnson, D.S.: Computers and Intractability: A Guide to the Theory of NP-Completeness. W. H. Freeman & Co., New York, NY, USA (1990).
12. Gollapudi, S., Sharma, A.: An axiomatic approach for result diversification. In: Proceedings of the International Conference on World Wide Web (WWW), pp. 381–390. ACM (2009). DOI http://dx.doi.org/10.1145/1526709.1526761

13. Hadjieleftheriou, M., Tsotras, V.J.: Letter from the special issue on result diversity. IEEE Data Eng. Bull. **32**(4), 6 (2009).

14. Hassin, R., Rubinstein, S., Tamir, A.: Approximation algorithms for maximum dispersion. Oper. Res. Lett. **21**(3), 133–137 (1997). DOI http://dx.doi.org/10.1016/S0167-6377(97)00034_5

15. Ioannou, E., Papapetrou, O., Skoutas, D., Nejdl, W.: Efficient semantic-aware detection of near duplicate resources. In: Proceedings of the Extended Semantic Web Conference (ESWC), LNCS, pp. 136–150. Springer (2010).

16. Jain, A., Sarda, P., Haritsa, J.: Providing diversity in k-nearest neighbor query results. In: H. Dai, R. Srikant, C. Zhang (eds.) Advances in Knowledge Discovery and Data Mining, Lecture Notes in Computer Science, vol. 3056, pp. 404–413. Springer, Berlin Heidelberg (2004). DOI http://dx.doi.org/10.1007/978-3-540-24775-3_49

17. Kuby, M.J.: Programming models for facility dispersion: The p-dispersion and maxisum dispersion problems. Geogr. Analysis **19**, 315–329 (1987). DOI http://dx.doi.org/10.1111/j.1538-4632.1987.tb00133.x

18. Kuby, M.J.: Programming models for facility dispersion: the p-dispersion and maxisum dispersion problems. Mathematical and Computer Modelling **10**(10), 792 - (1988). DOI http://dx.doi.org/10.1016/0895-7177(88)90094--5

19. Kuo, C.C., Glover, F., Dhir, K.S.: Analyzing and modeling the maximum diversity problem by zero-one programming. Decision Sciences **24**(6), 1171–1185 (1993). DOI http://dx.doi.org/10.1111/j.1540-5915.1993.tb00509.x

20. Laguna, M., Martí, R.: GRASP and path relinking for 2-layer straight line crossing minimization. INFORMS Journal on Computing **11**(1), 44–52 (1999).

21. van Leuken, R., Garcia, L., Olivares, X., van Zwol, R.: Visual diversification of image search results. In: Proceedings of the International Conference on World Wide Web (WWW), pp. 341–350. ACM (2009). DOI http://dx.doi.org/10.1145/1526709.1526756

22. Ley, M.: The DBLP Computer Science Bibliography. www.informatik.uni-trier.de/~ley/db

23. Liu, K., Terzi, E., Grandison, T.: Highlighting diverse concepts in documents. In: Proceedings of the SIAM International Conference on Data Mining (SDM), pp. 545–556. SIAM (2009)

24. Liu, Z., Sun, P., Chen, Y.: Structured search result differentiation. Proceedings of the VLDB Endowment (PVLDB) **2**(1), 313–324 (2009)

25. Prais, M., Ribeiro, C.: Reactive GRASP: An application to a matrix decomposition problem in TDMA traffic assignment. INFORMS Journal on Computing **12**(3), 164–176 (2000). DOI http://dx.doi.org/10.1287/ijoc.12.3.164.12639

26. Prokopyev, O., Kong, N., Martinez-Torres, D.: The equitable dispersion problem. European J. of Operational Research **197**(1), 59–67 (2009)

27. Radlinski, F., Dumais, S.: Improving personalized web search using result diversification. In: Proceedings of the ACM SIGIR International Conference on Research and Development in Information Retrieval, pp. 691–692. ACM (2006). DOI http://dx.doi.org/10.1145/1148170.1148320

28. Radlinski, F., Kleinberg, R., Joachims, T.: Learning diverse rankings with multi-armed bandits. In: Proceedings of the International Conference on Machine Learning (ICML), pp. 784–791. ACM (2008). DOI http://dx.doi.org/10.1145/1390156.1390255

29. Rafiei, D., Bharat, K., Shukla, A.: Diversifying web search results. In: Proceedings of the International Conference on World Wide Web (WWW), pp. 781–790. ACM (2010). DOI http://dx.doi.org/10.1145/1772690.1772770

30. Resende, M.G.C., Ribeiro, C.C.: Greedy randomized adaptive search procedures: Advances, hybridizations, and applications. In: M. Gendreau, J.Y. Potvin (eds.) Handbook of Metaheuristics, International Series in Operations Research & Management Science, vol. 146, 2 edn., pp. 283–320. Springer US (2010). DOI http://dx.doi.org/10.1007/978-1-4419-1665-5_10

31. Santos, R., Macdonald, C., Ounis, I.: Exploiting query reformulations for web search result diversification. In: Proceedings of the International Conference on World Wide Web (WWW), pp. 881–890. ACM (2010). DOI http://dx.doi.org/10.1145/1772690.1772780

32. Santos, R., Peng, J., Macdonald, C., Ounis, I.: Explicit search result diversification through sub-queries. In: C. Gurrin, Y. He, G. Kazai, U. Kruschwitz, S. Little, T. Roelleke, S. Rüger, K. Rijsbergen (eds.) Advances in Information Retrieval, Lecture Notes in Computer Science, vol. 5993, pp. 87–99. Springer, Berlin Heidelberg (2010). DOI http://dx.doi.org/10.1007/978-3-642-12275-0_11

33. Silva, G., de Andrade, M., Ochi, L., Martins, S., Plastino, A.: New heuristics for the maximum diversity problem. J. of Heuristics **13**(4), 315–336 (2007). DOI http://dx.doi.org/10.1007/s10732-007-9010-x

34. Smyth, B., McClave, P.: Similarity vs. diversity. In: Proceedings of the International Conference on Case-Based Reasoning: Case-Based Reasoning Research and Development (ICCBR), pp. 347–361 (2001)

35. The Text REtrieval Conference (TREC): Trec-3 collection. http://trec.nist.gov

36. Vee, E., Srivastava, U., Shanmugasundaram, J., Bhat, P., Amer-Yahia, S.: Efficient computation of diverse query results. In: Proceedings of the IEEE International Conference on Data Engineering (ICDE), pp. 228–236. IEEE Computer Society (2008). DOI http://dx.doi.org/10.1109/ICDE.2008.4497431

37. Yu, C., Lakshmanan, L., Amer-Yahia, S.: It takes variety to make a world: diversification in recommender systems. In: Proceedings of the International Conference on Extending Database Technology (EDBT), pp. 368–378. ACM (2009). DOI http://dx.doi.org/10.1145/1516360.1516404

38. Zhai, C., Cohen, W., Lafferty, J.: Beyond independent relevance: methods and evaluation metrics for subtopic retrieval. In: Proceedings of the ACM SIGIR International Conference on Research and Development in Information Retrieval, pp. 10–17. ACM (2003). DOI http://dx.doi.org/10.1145/860435.860440

39. Zhang, B., Li, H., Liu, Y., Ji, L., Xi, W., Fan, W., Chen, Z., Ma, W.Y.: Improving web search results using affinity graph. In: Proceedings of the ACM SIGIR International Conference on Research and Development in Information Retrieval, pp. 504–511. ACM (2005). DOI http://dx.doi.org/10.1145/1076034.1076120

40. Zheng, Y., Zhang, L., Xie, X., Ma, W.Y.: Mining interesting locations and travel sequences from GPS trajectories. In: Proceedings of the International Conference on World Wide Web (WWW), pp. 791–800. ACM (2009). DOI http://dx.doi.org/10.1145/1526709.1526816

41. Zhu, X., Goldberg, A.B., Gael, J.V., Andrzejewski, D.: Improving diversity in ranking using absorbing random walks. In: Proceedings of the Human Language Technology Conference of the North American Chapter of the Association of Computational Linguistics (NAACL-HLT), pp. 97–104 (2007)

42. Ziegler, C.N., McNee, S., Konstan, J., Lausen, G.: Improving recommendation lists through topic diversification. In: Proceedings of the International Conference on World Wide Web (WWW), pp. 22–32. ACM (2005). DOI http://dx.doi.org/10.1145/1060745.1060754

Chapter 6
Conclusion

This book presents several novel complex motion pattern queries for trajectory data. Previous works on querying trajectory data have mainly focused on traditional spatio-temporal queries, similarity/clustering based tasks, or spatio-temporal joins. Nevertheless, trajectories are complex objects whose behavior over space and time can be better captured as a *sequence* of interesting events, or the *aggregate* behavior of trajectories as groups. Given the deficiencies of previous approaches, this book describes several motion pattern queries that allow users to select trajectories based on specific events of interest.

This book introduces the *flexible pattern query*, a very powerful, yet easy to use motion pattern query which allows users to select trajectories based on specific events of interest. Such queries combine the ability of fixed and variable predicates, with explicit or implicit temporal constraints and distance-based constraints. Two query processing techniques are described: one based on merge joins (*IJP*) and one based on subsequence matching (*DPP*). The experimental evaluation shows that our techniques improve substantially even over *optimized* (using indexing and preprocessing techniques) previous approaches. Among all the described approaches, *IJP* is more robust in that it can easily support *NN* queries, while *DPP* is better for patterns with smaller number of predicates or wild-cards. Since, however, both approaches are implemented in the same FlexTrack system, they can both be available to the user.

Then, the Spatio-Temporal Pattern System (STPS) is proposed for processing spatio-temporal pattern queries over mobile phone-call databases. STPS defines a language to express pattern queries which combine fixed and variable spatial predicates with explicit and implicit temporal constraints. The STPS index structures and algorithm are described in order to efficiently process such pattern queries. The experimental evaluation shows that the STPS can answer spatio-temporal patterns efficiently even for very large mobile phone-call databases. Among the advantages of the STPS is that it can be easily integrated in commercial telecommunication databases and also be implemented in any current commercially available RDBMS.

M. R. Vieira and V. J. Tsotras, *Spatio-Temporal Databases*, 113
SpringerBriefs in Computer Science, DOI: 10.1007/978-3-319-02408-0_6,
© The Author(s) 2013

The next motion pattern query described in this book is the *flock pattern query*, which captures the collaborative behavior of spatio-temporal data. The result of the flock pattern query returns a group of at least μ trajectories which are all "enclosed" by a disk of diameter ε for at least δ consecutive timestamps. Discovering on-line flock patterns is useful for several applications, ranging from tracking suspicious activities to migrations of animals. Previous related approaches either do not work for on-line datasets, or do not return exact solutions for flock pattern queries. We first show that the discovery of flock patterns with fixed time duration can be computed in polynomial time. Then, a framework is presented that uses a lightweight grid-based structure in order to efficiently and incrementally process the trajectory's locations. Experiments on various trajectory datasets show that the proposed methods can efficiently report flock patterns even for large datasets and for different variations of the flock parameters (μ, ε and δ).

Finally, this book presents the DivDB framework, a simple yet powerful framework for evaluation and optimization of methods for diversifying query results. One advantage of the DivDB framework is that users can set the *tradeoff* between finding the most relevant elements to the query and finding diverse elements in the result set. Another advantage is that one can quantitatively compare different methods implemented in the DivDB framework. Then, several known methods as well as two new methods for diversifying query results are described. The two new proposed methods, named GMC and GNE, construct the result set in an incremental way using a function that computes the contribution of each candidate element considering its relevance to the query and its diverse value to the elements in the current result set and to other elements in the candidate set. In GMC, the element with the highest contribution value is always included in the result set, while in GNE, a randomized procedure is employed to choose, among the top highest elements, the one to be included in the result set. A thorough experimental evaluation of the various diversification methods using real datasets shows that our new proposed methods implemented in the DivDB framework give better results than previous approaches.